Toxic Exports

Toxic Exports

The Transfer of Hazardous Wastes from Rich to Poor Countries

Jennifer Clapp

Cornell University Press
Ithaca and London

Copyright © 2001 by Cornell University

All rights reserved. Except for brief quotations in a review, this book, or parts thereof, must not be reproduced in any form without permission in writing from the publisher. For information, address Cornell University Press, Sage House, 512 East State Street, Ithaca, New York 14850.

First published 2001 by Cornell University Press

Printed in the United States of America

Library of Congress Cataloging-in-Publication Data

Clapp, Jennifer, 1963–
    Toxic exports : the transfer of hazardous wastes from rich to poor countries / Jennifer Clapp.
      p. cm.
    Includes bibliographical references and index.
    ISBN 0-8014-3887-X (cloth : alk. paper)
    1. Hazardous wastes—Developing countries. 2. Hazardous wastes—Transportation. 3. Globalization. I. Title.
    TD1045.D44 C58 2001
    363.72'87'091724—dc21   2001001649

Cornell University Press strives to use environmentally responsible suppliers and materials to the fullest extent possible in the publishing of its books. Such materials include vegetable-based, low-VOC inks and acid-free papers that are recycled, totally chlorine-free, or partly composed of nonwood fibers. Books that bear the logo of the FSC (Forest Stewardship Council) use paper taken from forests that have been inspected and certified as meeting the highest standards for environmental and social responsibility. For further information, visit our website at www.cornellpress.cornell.edu.

Cloth printing   10 9 8 7 6 5 4 3 2 1

For Eric, Zoë, and Nels

Contents

List of Tables and Boxes   viii

Preface   ix

List of Acronyms   xi

1 Hazard Transfer from Rich to Poor Countries   1

2 The Hazardous Waste Trade and International Regulatory Measures   21

3 The Role of Environmental NGOs in the Evolution of the Basel Ban   53

4 Industry Players and Post–Basel Ban Amendment Politics   81

5 Foreign Direct Investment in Hazardous Industries   104

6 Market-Based and Voluntary Initiatives: Promoting Clean Production?   126

7 Conclusion: Prospects for Clean Production on a Global Scale   150

Index   173

Tables

2.1   Generation of Hazardous Wastes in Select OECD Countries   25

2.2   Summary of Transfrontier Movements of Hazardous Wastes from 1989 to 1993 in OECD Countries   28

2.3   Number of OECD to Non-OECD Waste Trade Schemes by Year   30

2.4   Number of Schemes Proposed for Exports by Receiving Region and Year   33

2.5   Results of Hazardous Waste Trade Proposals from OECD to Non-OECD Countries, 1989–1993   34

Boxes

2.1   The Health Impacts of Toxic Wastes: Some Examples   26

3.1   Some Examples of Exports of Hazardous Waste from OECD to Non-OECD Countries Destined for Recycling Operations, 1995–1997   59

3.2   Text of Decision II/12   74

3.3   Text of Decision III/1   78

Preface

In this book I examine the transfer of hazardous wastes and technologies from rich to poor countries. I look at forces that contribute to that transfer, as well as the political responses to it. The phenomenon is a product of economic globalization in the context of a highly unequal world, and it has generated various political responses, ranging from efforts to halt exports of toxic waste to calls for the transfer of cleaner production technologies. These initiatives all have serious weaknesses, or are under threat of being weakened. The reason, I suggest, is that hazard transfer is both dynamic and multifaceted. Efforts to stop one form of toxic exports prompt new forms to emerge. The players in this cat and mouse game are not just states and the specific corporations they seek to regulate. Also important are increasingly powerful nongovernmental organizations and industry lobby groups operating at the international level.

This project had its origins in my travels in West Africa in the late 1980s, when the region was the recipient of numerous shipments of hazardous waste from rich countries. My early research into the issue quickly made clear to me that these shipments were part of a broader trend. I am grateful to Gwyn Prins and the MacArthur Foundation for giving me the chance to explore this issue in depth, on a postdoctoral fellowship at the University of Cambridge. Prins's guidance and enthusiasm for the project helped launch what was initially a relatively small research program into a much larger and longer-term study.

In subsequent years, many other people contributed to this project in ways that I would like to acknowledge. I owe my deepest gratitude to Eric Helleiner, for his untiring intellectual and emotional support throughout. Special thanks are also due to those who took time out of their busy schedules to speak with me, including Harvey Alter, Julie Gourley, Scott Horne, Janice Jensen, Katharina Kummer, Klaus Lingner, Pierre Portas, Jim Puckett, Mary Clock Rust, Ellen Spitalnik, Kevin Stairs, and Jim Vallette. Jim Puckett in particular was extremely generous with his time, and provided helpful comments on last-minute drafts. I also thank a number of colleagues for their feedback and support on various papers presented at conferences of the International Studies Association and the American Political Science Association. They are Ken Conca, Peter Dauvergne,

Matthias Finger, Virginia Haufler, Jonathan Krueger, David Levy, Ronnie Lipschutz, Marian Miller, Kate O'Neill, Tom Princen, Gene Rochlin, Ian Rowlands, Paul Wapner, Marc Williams, and Mark Zacher. Two anonymous reviewers deserve thanks for their insightful comments on earlier drafts. I also owe thanks to an extremely impressive group of research assistants who have helped with this project. They are Matt Griem, Christopher Mahood, Shawn Morton, Nancy Palardy, Rolie Srivastava, Eric Verreault, Ken Watt, David Wallbridge, and Jacob Wilson. I am grateful to Roger Haydon for shepherding the manuscript through the publication process. The generous financial support of the Social Sciences and Humanities Research Council of Canada made this project possible. Finally, thanks are due to my family, friends, and colleagues for their patience and support, particularly in the final stages of this project.

JENNIFER CLAPP

Peterborough, Ontario

Acronyms

ACP	African, Caribbean, and Pacific (nations)
ASEAN	Association of Southeast Asian Nations
BAN	Basel Action Network
BIR	Bureau of International Recycling
BRC	Business Recycling Council
CAC	command and control
CARICOM	Caribbean Community
CFCs	chlorofluorocarbons
COP	Conference of the parties
DTIE	Division of Trade, Industry, and Economics (UNEP)
EC	European Community
ECLAC	Economic Community of Latin American Countries
ECOWAS	Economic Community of West African States
EMAS	Environmental Management and Audit Scheme (EU)
EMS	environmental management system
EPA	Environmental Protection Agency (U.S.)
ESM	environmentally sound management
EU	European Union
FDI	foreign direct investment
FPG	Formosa Plastics Group
G-77	Group of 77 (developing countries)
GATT	General Agreement on Tariffs and Trade
GDP	gross domestic product
GNP	gross national product
HCB	hexachlorobenzine
IAEA	International Atomic Energy Agency
ICC	International Chamber of Commerce
ICME	International Council on Metals and the Environment
IMLI	Indo Era Multa Logam (battery recycling plant)
IPMI	International Precious Metals Institute
ISO	International Organization for Standardization
ISRI	Institute of Scrap Recycling Industries
ITWAN	International Toxic Waste Action Network
MEA	multilateral environmental agreement

NAFTA	North American Free Trade Agreement
NAM	Non-Aligned Movement
NGO	nongovernmental organization
OAU	Organization of African Unity
OECD	Organization for Economic Cooperation and Development
PCBs	polychlorinated biphenols
PRI	Philippine Recyclers Incorporated
PRTR	pollution release and transfer register
PVC	polyvinyl chloride
SAGE	Strategic Advisory Group on the Environment
SMEs	small- and medium-scale enterprises
TBT	technical barriers to trade
TNC	transnational corporation
TWG	Technical Working Group
UNCED	United Nations Conference on Environment and Development
UNCTAD	United Nations Conference on Trade and Development
UNCTC	United Nations Centre on Transnational Corporations
UNEP	United Nations Environment Programme
USAID	United States Agency for International Development
WBCSD	World Business Council on Sustainable Development
WHO	World Health Organization
WTO	World Trade Organization

Toxic Exports

1
Hazard Transfer from Rich to Poor Countries

Just between you and me, shouldn't the World Bank be encouraging *more* migration of the dirty industries to the LDCs [less developed countries]?... I think the economic logic behind dumping a load of toxic waste in the lowest wage country is impeccable and we should face up to that... I've always thought that under-populated countries in Africa are vastly *under*-polluted.

—Lawrence Summers, 1991, then chief economist at the World Bank, excerpts from leaked internal memo[1]

Larry Summers's memo caused quite a stir. It was written to colleagues while he was leading a team of economists in drafting a major report on development and the environment. The memo provoked strong opposition among the general public to the idea of promoting a transfer of hazards from richer to poorer countries. After the leak to the press in early 1992, Summers felt the need to apologize. He said that his comments had been only tongue-in-cheek. Transfers of the most hazardous wastes and technologies from rich to poor countries may be "perfectly logical" in an economic sense, but many observers see them as "totally

---

[1] This memo was reprinted in *The Economist*, February 8, 1992, p. 66. Summers would later become U.S. treasury secretary.

insane," José Lutzenberger, then secretary of the environment in Brazil, told Summers.[2] This sentiment has been the basis for international efforts to control the waste trade since the mid-1980s. But while there is widespread agreement, at least on the surface, that transfers of hazardous wastes and industries from richer to poorer countries should not be sanctioned, they persist in various forms. Why is this so?

Much of the international trade in hazards takes place among rich industrialized countries, where the wastes originated.[3] But a significant portion of hazards have found their way to less industrialized countries. Notorious cases in the late 1980s, such as the Italian toxic waste in Nigeria and the voyage of the *Khian Sea*, which attempted to dump toxic waste of U.S. origin in several developing countries, are now the stuff of legend. Other forms of hazard transfer have since emerged, among them the export from industrialized countries of toxic wastes destined for recycling operations in the developing world. Another troubling trend is foreign direct investment in hazardous manufacturing facilities, using outdated equipment and techniques. Because they lack financial resources, poorer countries are unlikely to be able to manage hazards in ways that protect the environment and human health. The transfer of hazards is also delaying the adoption of clean production in both rich and poor countries.

Economic globalization is related to this transfer of hazards from richer to poorer countries because global networks for trade and investment facilitate the relocation of hazards. But as the public response to Summers's memo indicates, the global community has not accepted this practice as an unavoidable consequence of globalization. Wider recognition of the hazard transfer problem has led to international cooperation that ostensibly aims to halt it. The countertrends include international agreements such as the Basel Convention and industry efforts to clean up their acts. Nonstate actors, including environmental nongovernmental organizations (NGOs) and business lobby groups, have been extremely influential in shaping these countertrends.

If globalization encourages environmental problems but also leads to countertrends that address them, optimists might expect the system to be self-correcting. Unfortunately, the problem is much more complex. International regulation may put a damper on certain forms of hazard transfer,

---

[2] Quoted in Jim Vallette, "The Tragic Rise of the New Treasury Secretary," *International Trade Information Service*, May 13, 1999.
[3] Kate O'Neill, *Waste Trading among Rich Nations: Building a New Theory of Environmental Regulation* (Cambridge, Mass.: MIT Press, 2000), offers an excellent analysis of this trade among OECD countries.

but the practice finds new outlets in response. Various types of hazard transfer, it turns out, are related, and the problem is a dynamic one. Plugging one hole in the dike tends to create a new one elsewhere.

A brief history of the problem reveals its dynamic nature. In the 1980s it became apparent that hazardous wastes generated in industrialized countries were being shipped to developing countries for final disposal. Cost differentials between rich and poor countries were too attractive for waste dealers to ignore, and global trade and communications networks made this disposal feasible. The practice affected countries in nearly every region of the developing world. When environmental NGOs and the media first publicized the practice, many saw it as morally wrong and demanded international action. In 1989, in response to the growth of the waste trade, states came together to negotiate and sign the Basel Convention on the Transboundary Movement of Hazardous Wastes and Their Disposal. Agreement on this treaty was relatively swift, as both industrialized and less industrialized countries acknowledged that hazardous wastes should not be subject to free trade. There was much disagreement, however, over whether the practice should be outlawed entirely. The end result was a treaty that sought to control, rather than ban, the trade in hazardous waste. Following an NGO campaign in the developing world, regional and national laws also emerged to ban imports of waste.

By the mid-1990s, the export of toxic wastes for final disposal in developing countries had slowed down markedly, in large part as a result of the various regulations and negative media attention. But the problem of hazard transfer did not disappear. Rather, it evolved. A growing number of firms in richer countries began to export toxic waste to recycling operations in poorer countries. Wastes destined for recycling operations are technically covered by the Basel Convention. But there is a loophole: when wastes are not labeled as wastes for recycling, they are difficult to regulate. Though recycling infers environmental stewardship over these wastes, in most cases the recycling of imported hazardous waste in developing countries has proved just as harmful as outright disposal. Environmental NGOs called attention to this new phenomenon and led efforts to stop it. Developing countries were largely in agreement that the Basel Convention should incorporate explicit measures to address hazard transfer via recycling. But several key Organization for Economic Cooperation and Development (OECD) countries, as well as increasingly vocal industry lobby groups, put up a fight. A heated debate lasted several years, and the global community finally adopted, in 1995, an amendment to the Basel Convention that bans the export of toxic wastes from OECD to non-OECD countries for both disposal *and* recycling. This amendment represents a pre-emptive move on the part of the global community to avoid a major waste transfer crisis before it became full-blown.

The adoption of the Basel Ban Amendment brought a sense of victory to environmental NGOs and states fighting against international transfers of toxic waste. The amendment requires ratification by 62 parties to the convention before it comes into effect, however, and this may take some time. Though the ban amendment is not yet in force, the fact that it was adopted by the parties to the Basel Convention has in practice substantially reduced the transfer of hazardous waste, both for disposal and recycling, between rich and poor countries. It has not completely ended the transfer of hazardous wastes, which is why environmental NGOs argue that it is important that the ban amendment be ratified. But there are forces working against this goal. Immediately after the adoption of the ban amendment in 1995, a further aspect of the hazard transfer problem became apparent. The global recycling industry began to try to weaken the ban. They tried to redefine hazardous waste in the context of the Basel Convention, to ensure that its business in these wastes was not harmed by the ban. Moreover, these groups also began trying to reverse the Basel ban in an attempt to pave a legal channel for the transfer of recyclable hazardous waste.

Even if the Basel Ban Amendment does come into force soon, some environmental groups fear that it will only accelerate an existing and potentially more damaging form of hazard transfer: the migration of entire hazardous waste–generating industries and production processes to developing countries. The question of whether industries migrate for environmental reasons has been debated in recent decades. But there is much documented evidence that the *most* hazardous activities of multinational firms have *already* relocated to developing countries. Environmental NGOs have campaigned against such transfers of hazardous industry. If the waste trade for both disposal and recycling is banned, NGOs expect that hazardous technology transfer will grow in the future, particularly in the newly industrializing countries of Latin America and Asia, as well as in Eastern Europe. For this reason they are now calling for measures to clean up global industry.

The transfer of clean production technologies from richer to poorer countries is widely perceived to be the most promising solution to both the problems of hazardous waste disposal and the migration of hazardous industry. Even business associations now argue that transnational corporations (TNCs) will recognize that it is more economically efficient to transfer clean production processes to developing countries than it is to continue with dirty or hazardous practices. TNC representatives promised at the Rio Earth Summit to enhance their efforts to transfer cleaner technologies to developing countries to rectify this problem. But the result in many instances has been a transfer of cleanup technologies to deal with existing hazardous wastes rather than a shift to cleaner production technologies. The problem has taken on yet another form.

Local and international environmental NGOs, as well as business organizations and multinational firms, are increasingly recognizing this problem. This time TNCs and industry lobby groups have dominated the political response. Efforts are being made to address the problem through voluntary industry standards for environmental management, such as ISO 14000, which seek to create incentives to install clean production technologies globally.[4] This may be a preemptive move by industry players to capture the control of market and voluntary initiatives so as to maintain their ability to transfer hazards globally while at the same time appearing to be "green." In emphasizing these voluntary measures, industry players are able to avoid more stringent regulation of technology through legally binding instruments. But it is not at all clear that the ISO 14000 standards will help to transfer cleaner technology to developing countries or reduce hazardous waste generation. This uncertainty has prompted environmental NGOs and other critics to challenge the usefulness of these measures.

This history shows that hazards have been transferred from rich to poor countries in today's global economy and that various political countertrends that ostensibly aim to regulate the hazard transfer problem have also emerged. But that is not all. There have also been responses to the political countertrends by those in toxic waste–generating industries and waste disposal firms, as well as by lobby groups representing these industries. The global economy has facilitated the ability of these groups to respond in ways that allow hazards to be transferred via new outlets that are as yet unregulated. The result is a global cat and mouse game of hazard transfer.

The global environmental politics literature offers useful insights into some aspects of this story. The two subfields that should have the most to say about this problem are the trade and environment literature, on the one

---

[4] Charles Hadlock, "Multinational Corporations and the Transfer of Environmental Technology to Developing Countries," *International Environmental Affairs* 6, no. 2 (1994): 149–74; Naomi Roht-Arriaza, "Shifting the Point of Regulation: The International Organization for Standardization and Global Lawmaking on Trade and the Environment," *Ecology Law Quarterly* 22, no. 3 (1995): 479–539.

hand, and the nonstate actor literature, on the other. But the former has paid insufficient attention to the features of the global economy that facilitate the continual evolution of the hazard transfer problem. And the latter has not paid adequate attention to the role of business actors in global environmental policy-making. Moreover, while the broader global environmental politics literature has tended to see the trade and environment issue as somewhat separate from that of nonstate actors, the case of rich-to-poor country hazard transfer demonstrates that they are closely linked.

The Global Economy and Hazard Transfer

There has been a great deal of discussion in recent years on the interface between the operation of the global economy—particularly trade and investment—and the natural environment. A heated debate among policymakers, economists, and activists has emerged and is polarized by two very separate views. This debate encompasses broader discussions on the environmental implications of economic growth that may result from global economic integration. It also includes more specific discussions regarding global economic relationships and the environment. Because this debate is explained in detail by others, I will provide only a brief overview of it here and discuss its relevance to the issue of hazard transfer.[5]

Liberal economists have tended to dominate one side of these debates. They argue that the liberalization of trade, investment, and financial rules—what many say is the driving force behind economic "globalization"—is beneficial to the natural environment because it encourages economic growth. Growth causes rising incomes, which encourage two things. First, higher incomes are associated with a higher demand for a cleaner environment. Second, rising incomes mean that more economic resources are available which can then be spent on environmental protection.[6] But before this stage is reached, poorer countries are seen to have a higher capacity to absorb pollution, which explains their less stringent environmental regulations. It is also considered to be a part of their comparative advantage.[7]

---

[5] For an overview of the debate, see, for example, Daniel Esty, *Greening the GATT* (Washington, D.C.: IIE, 1994); Marc Williams, "International Trade and the Environment: Issues, Perspectives and Challenges," in *Rio: Unraveling the Consequences*, ed. Caroline Thomas (Ilford: Frank Cass, 1994), 80–97.

[6] Gene Grossman and Alan Krueger, "Economic Growth and the Environment," *Quarterly Journal of Economics* 40 (1995): 353–77.

[7] For an overview, see Gareth Porter, "Pollution Standards and Trade: The 'Environmental Assimilative Capacity' Argument," *Georgetown Public Policy Review* 4, no. 1 (1998): 50–52.

Those coming from this dominant perspective argue that despite poor countries' advantage in pollution absorption, TNCs are unlikely to relocate in developing countries in order to take advantage of lax environmental laws.[8] In other words, industry does not take "flight" to developing countries in response to more stringent regulations at home, and thus developing countries are not "pollution havens." Rather, they argue, TNCs that do set up shop in developing countries tend to be more environmentally sound than their local counterparts.[9] It is argued that it is in TNCs' best economic interests to go green. The "race to the bottom" phenomenon, through which countries lower environmental regulations to gain competitiveness, is seen by this view to be highly unlikely. Instead, the global economy encourages an upward movement of environmental regulations, or a "race to the top."[10] Following this line of reasoning, many have argued that free trade agreements should take precedence over multilateral environmental agreements (MEAs). Trade restrictions contained in MEAs are seen to constitute potential trade barriers that will have negative implications both for economic growth and for the environment.[11]

Though this dominant view among liberal economists has received a great deal of attention in policy circles, there has been a growing chorus of opposition. Environmentalists, activists, and ecological economists have argued that economic liberalization and globalization are at the root of environmental destruction around the globe today. A key argument these groups make is that free trade, while it may bring economic growth, has also increased physical throughput in the economy. This is the case even after taking into account any increases in efficiency that may have been gained through technology improvements.[12] Moreover, some have argued

---

[8] Patrick Low and Alexander Yeats, "Do 'Dirty' Industries Migrate?" in *International Trade and the Environment*, ed. Low (Washington, D.C.: World Bank, 1992), 89–103. Muthukumara Mani and David Wheeler, "In Search of Pollution Havens? Dirty Industry in the World Economy, 1960 to 1995," *Journal of Environment and Development* 7, no. 3 (1998): 215–47.

[9] Norman Bailey, "Foreign Direct Investment and Environmental Protection in the Third World," in *Trade and the Environment*, ed. Durwood Zaelke et al. (Washington, D.C.: Island Press, 1993), 136.

[10] See, for example, with respect to the mining industry, Gordon Clark, "Global Competition and Environmental Regulation: Is the Race to the Bottom Inevitable?" in *Markets, the State and the Environment: Towards Integration*, ed. Robyn Ekersly (South Melbourne: Macmillan, 1995).

[11] Jagdish Bhagwati, "The Case for Free Trade," *Scientific American*, November 1993: 42–49.

[12] See, for example, Herman Daly, *Beyond Growth* (Boston: Beacon, 1996); Wolfgang Sachs, "Global Ecology and the Shadow of Development," in *Global Ecology*, ed. Sachs (London: Zed, 1993), 3–21.

that the liberal economic theory which claims that environmental improvements will be enjoyed as economies' incomes rise is flawed. A key reason is that this relationship is based on the experience of the already industrialized countries and may not apply to less industrialized countries in the current global economy.[13]

From this alternative perspective, trade and investment liberalization is also seen to give extraordinary powers to TNCs. This power encourages a regulatory race to the bottom, as TNCs can threaten to leave jurisdictions that do not conform with corporate demands for less environmental regulation.[14] From this perspective, industry flight and pollution havens are real threats. For these reasons, it is argued that international trade and investment need to be reined in with strong global-level regulatory measures to protect the environment. These measures should also incorporate assistance for developing countries to help them avoid environmental mistakes made by industrialized countries. Following this line of reasoning, many from this perspective argue that trade restrictions for environmental reasons, whether in MEAs or imposed by states unilaterally, are beneficial and necessary. But they are also skeptical of the ability of multilateral agreements alone to protect the environment. Thus these thinkers also call for a new human ethic based not on global competition and economic growth but rather on community development at the local level.[15]

Within this broader debate, three specific strands of inquiry have direct relevance to the problem of hazard transfer. These are, first, the impact of environmental regulations on countries' trade competitiveness. Second is the role environmental regulations play in industrial location. And third is the compatibility of trade rules with trade measures incorporated into multilateral environmental agreements. Each of these bodies of literature, on both sides of the debate, has important insights to add to our understanding of the hazard transfer problem. But as I explain below, they do not adequately probe its dynamic nature.

The debate over whether a country's international economic competitiveness is enhanced by relaxing its environmental regulations is important

---

[13] Kenneth Arrow et al., "Economic Growth, Carrying Capacity and the Environment," in *Debating the Earth*, ed. John Dryzek and David Schlosberg (Oxford: Oxford University Press, 1998), 35–39.

[14] See, for example, Joshua Karliner, *The Corporate Planet* (San Francisco: Sierra Club, 1997); David Korten, *When Corporations Rule the World* (West Hartford, Conn.: Kumarian Press, 1995); Pratap Chatterjee and Matthias Finger, *The Earth Brokers* (London: Routledge, 1994).

[15] See, for example, numerous chapters in Edward Goldsmith, ed., *The Case against the Global Economy* (San Francisco: Sierra Club, 1996).

for the hazard transfer question. Relaxed regulations with respect to hazardous waste management might make a country a favored destination for the import of waste, and would earn the country foreign exchange. Tighter international environmental agreements that call for national regulatory strengthening in one area might lead countries to relax regulations in another area in order to improve competitiveness. But literature on this topic has tended to focus almost exclusively on the impact of domestic environmental regulations on countries' export performance.[16] It pays little attention to the impact of weak domestic regulations on the inward movement of hazards such as waste imports. This is partly because much of the literature on this question is focused on the implications of a potential race to the bottom for environmental quality in rich industrialized countries, rather than its potential impact on developing countries. The literature also fails to analyze the response of firms to more stringent international, as opposed to domestic, environmental regulations. This is an important distinction in the case of hazards because focusing only on firms' response to domestic regulations fails to capture the broader response to the tightening of international rules such as the Basel Convention.

The related debate on industry flight and pollution havens has important implications for the question of hazard transfer. If firms do in fact relocate to developing countries in order to take advantage of relatively more relaxed environmental regulations, we would expect to see increased foreign direct investment in hazardous industries in those countries. Interestingly, most of the literature on this topic acknowledges that a transfer of the most hazardous industries from rich to poor countries has occurred as a response to more stringent environmental regulations in industrialized countries. But this phenomenon is listed in most studies as an exception to the general trend, which is that the broader category of "polluting" industries generally do not relocate for environmental reasons.[17] There is very little discussion in this literature of why this exception for the most hazardous industries continues to exist or what should be done about it. There is also little connection made between the industry location issue and the

---

[16] Candice Stevens, "Do Environmental Policies Affect Competitiveness?" *OECD Observer*, no. 183 (August–September 1993): 22–25; Richard Stewart, "Environmental Regulation and International Competitiveness," *Yale Law Journal* 102, no. 8 (1993): 2039–2106; Cees van Beers and Jeroen C. J. M. van den Bergh, "An Empirical Multi-Country Analysis of the Impact of Environmental Regulations on Foreign Trade Flows," *Kyklos* 50, Fasc. 1 (1997): 29–46.
[17] See, for example, H. Jeffrey Leonard, *Pollution and the Struggle for the World Product* (Cambridge: Cambridge University Press, 1988), 111.

increasing number of regulations on the international trade in wastes, at both the domestic and international levels. This is important because as regulations on the hazardous waste trade become more strict, firms may seek to relocate entire manufacturing plants that produce those wastes in significant quantities.

Several studies have explored the compatibility of trade agreements and environmental agreements, some of which deal directly with the hazard transfer issue. Considerable debate has emerged as to whether the trade restrictions incorporated in the Basel Convention, in particular the Basel Ban Amendment, contravene global trade rules as set by the World Trade Organization (WTO).[18] The literature on this subject is extremely important in laying out the legal aspects of attempting to regulate global transfers of hazards. But this debate tends to focus on legal issues and does not analyze the dynamic nature of the global liberal trade order, now enforced by the WTO, that gave rise to the hazard transfer problem in the first place.

None of these debates fully captures the importance of the increasingly global nature of the world economy as a key factor in hazard transfer. The closest they come is to point out the role of local incentives. In the industrialized countries, increased environmental concerns have brought rising disposal costs for hazardous wastes, creating an incentive to export wastes and hazardous manufacturing facilities to countries with lower disposal costs. In developing countries, a weak capacity for regulation and monitoring has resulted in little if any control on the disposal of hazardous wastes. But these factors, while important, are not alone sufficient to explain the existence of hazard displacement from rich to poor countries. The transfer could not easily happen were it not for the globalization of the world economy. Economic globalization has created a setting in which hazards escape regulations on a global scale and their transfer takes advantage of economic inequalities between countries. It is not a case of a race to the top or to the bottom but rather, a problem of entrenched regulatory differences and their exploitation through global economic channels. Gareth Porter has identified part of this problem in his investigation of what he calls the "stuck at the bottom" phenomenon among rapidly indus-

---

[18] Jonathan Krueger, *International Trade and the Basel Convention* (London: Earthscan, 1998); David Wirth, "International Trade in Wastes: Trade Implications of the Recent Amendment to the Basel Convention Banning North-South Trade in Hazardous Wastes," draft report, January 19, 1996; Maria Isolda P. Guevara and Michael Hart, *Trade Policy Implications of the Basel Convention Export Ban on Recyclables from Developed to Developing Countries* (Ottawa: International Council on Metals and the Environment, 1996).

trializing developing countries.[19] So what is it about the global economy that facilitates the dynamic response of hazards to new regulations on a global scale? Three key aspects of the global economy play a role.

Rapidly growing levels of international debt over the past two decades have increased the vulnerability of poor countries to global economic trends. Many developing countries implemented policies of structural adjustment under the tutelage of the IMF and World Bank in the 1980s and 1990s in return for the rescheduling of some of this debt by donor countries and banks. These policies generally called for the liberalization of trade and investment policies in adjusting countries. This situation, combined with the not unrelated domestic political and institutional weakness in many poor countries, has made developing countries ideal targets for rich countries' unwanted hazards because they came with a promise of much needed foreign exchange. But these countries were not begging for hazards to be transferred. Other aspects of the global economy enabled hazard traders to take advantage of developing countries' vulnerable economic position.

The increased fluidity of trade in today's global marketplace has been a particularly important channel for the movement of hazardous wastes. Lower transportation and communication costs, the relative ease with which trade routes are established and abandoned, and the difficulties involved in checking every import container have facilitated the transfer of hazards. This increased fluidity of trade has emerged as states, rich and poor alike, have adopted more liberal trade policies over the past twenty years. These qualities of global trade have made the export of hazardous waste to less industrialized countries a simple and lucrative business for waste entrepreneurs, both legal and illegal. As wastes have been increasingly disguised as other products or are sent abroad for recycling, detection of these shipments has become more difficult.

The globalization of the production process and the footloose nature of transnational investment have also facilitated the movement of hazards around the world. The globalization of trade has enabled firms to sell their products on the global market, regardless of where they are produced. As a result, investment has become increasingly global to take advantage of cost differentials. Growing levels of investment by TNCs since the 1970s have been connected with the wholesale migration of hazardous industry to poor countries, particularly as environmental regulations in industrialized

---

[19] Gareth Porter, "Trade Competition and Pollution Standards: 'Race to the Bottom' or 'Stuck at the Bottom'?" *Journal of Environment and Development* 8, no. 2 (1999): 133–51.

countries became more stringent. Firms that have experienced serious decline in rich industrialized countries or have been subjected to severe and costly environmental and health regulations have a history of moving to poorer and less regulated countries. The developing world's share of inward investment from abroad in pollution-intensive industries has been rising in the past two decades, while at the same time investment in these industries elsewhere has declined.[20] Though this pattern of foreign direct investment (FDI) in highly polluting industries has affected both developed and developing countries, the latter are much more vulnerable to this type of investment because they have weaker environmental regulations and/or lack of enforcement of such regulations. The liberalization of investment regulations in countries pursuing structural adjustment policies has played a role in opening up these countries to new investment of this sort.

Given these factors in the global political economy, new regulatory measures can easily be circumvented by hazardous waste-producing industries. The problem of hazard transfer or displacement then reappears but as a new problem, in another form. Though many see the issues of hazardous waste exports, multinational investment in dirty industries, and barriers to clean technology transfer as separate issues that require independent consideration, they are closely related. The ease with which hazards move in a fluid global economy and the low monetary costs associated with their displacement to poor countries have contributed to a situation in which the real costs of such industrial activity are severely undervalued and environmental hazards persist. The problem is merely being displaced to countries that have difficulty containing the environmental impact. The result is that there is little incentive for firms to transfer clean production technologies when it is much cheaper, at least in the short run, to transfer hazards.

Nonstate Actors and the Political Response to Hazard Transfer

The study of nonstate actors has been a growth area in international relations broadly and in global environmental politics in particular over the past decade. Emphasis on nonstate actors arose mainly out of the realization by many international relations scholars at the end of the Cold War

---

[20] United Nations Transnational Corporations and Management Division, Department of Economic and Social Development, *World Investment Report, 1992* (New York: United Nations, 1992), 231.

that a focus on states alone did not adequately reflect the reality of transnational politics.[21] Most of the attention in the global environmental politics literature that has centered on nonstate actors has been on international environmental NGOs, mainly because these actors were highly visible in international environmental negotiations in the 1990s. At the United Nations Conference on Environment and Development (UNCED) they participated in unprecedented numbers and gained legitimacy as important players in international environmental politics.[22] Important contributions have been made by scholars writing on NGOs, particularly with respect to the reasons behind the recent rise in the numbers of these actors, their influence on states in global environmental policy-making, and their broader role beyond interstate politics in raising peoples' consciousness regarding global environmental issues.

Several explanations for the recent surge in numbers and influence of nonstate actors, particularly environmental NGOs, have been put forward. Some have linked the rise of non-state actors in global environmental politics to the process of economic globalization. Ronnie Lipschutz, for example, argues that a "global civil society" of interconnected and likeminded NGOs working on various issue areas has emerged as a reaction to economic globalization. That is, the global nature of the world economy, which has been characterized by the hegemony of liberal economic values, has prompted individuals to resist it. They joined the campaigns of NGOs with both local and global agendas that are attempting to regain some control over the economic forces that govern their everyday lives.[23] The spread of technology that has helped to foster global NGO linkages has both encouraged and been driven by the globalization of the world economy. Others have disagreed as to the extent to which globalization is linked to the rise of these actors. Margaret Keck and Kathryn Sikkink argue, for example, that transnational advocacy networks are based more on compassion and morality than simply on a response to globalization.[24] Both of these explanations are useful in understanding the emergence of environmental NGOs that are seeking to halt hazard transfer. Clearly

---

[21] James Rosenau, *Turbulence in World Politics* (Princeton: Princeton University Press, 1990); Jessica Tuchman Mathews, "Power Shift," *Foreign Affairs* 76, no. 1 (1997): 50–66.
[22] Tom Princen and Matthias Finger, *Environmental NGOs in World Politics* (New York: Routledge, 1994).
[23] Ronnie Lipschutz, "Reconstructing World Politics: The Emergence of Global Civil Society," *Millennium* 21, no. 3 (1992): 389–420.
[24] Margaret Keck and Kathryn Sikkink, *Activists beyond Borders* (Ithaca: Cornell University Press, 1998), 14.

these groups are reacting to the way in which the current global economy is facilitating hazard transfer. But they also make a key point about the injustice of the practice. It is important also to point out that the issue of hazard transfer highlights the dynamic relationship between economic globalization and global civil society. These groups are not just a one-time reaction to globalization. As the protests at the Seattle WTO meeting in 1999 and other anti-globalization protests since then have shown, NGOs are tracking globalization, moving with it, and reacting to it, just as it reacts to them.

Other studies on the role of NGOs in global environmental politics have focused on their key position as diplomatic actors. Environmental NGOs have been increasingly recognized by states as legitimate players in most stages of the international environmental treaty process.[25] They have been key in identifying issues that require action, as well as in the negotiation, monitoring, and enforcing of international environmental treaties. One of the key factors behind this legitimacy is the special knowledge and expertise on environmental issues that these groups possess.[26] These groups are very active in the negotiating sessions of environmental agreements. They are often key advisers on strategy to certain state decision makers behind the scenes, sometimes even participating on state delegations. They also have taken a key role in ensuring that states live up to their treaty obligations by publicizing their performance. Their role in monitoring compliance is extremely important, as it is a task that is easier for these groups to undertake than for treaty secretariats, who are more subject to state influence and take great pains to maintain neutrality. In the case of hazard transfer, NGOs played a key role in the identification of the waste trade and in the negotiation and monitoring of global and regional waste trade agreements. Although NGOs have been major actors in setting the tone and terms of the international debate on toxic transfer, surprisingly little has been written on their role in it. This neglect represents a failure to recognize their strong influence in setting the terms of the debates and drafting international treaties on the waste trade.

---

[25] Thomas Princen, "NGOs: Creating a Niche in Environmental Diplomacy," in *Environmental NGOs in World Politics: Linking the Local and the Global*, ed. Thomas Princen and Matthias Finger (London: Routledge, 1994).

[26] Steve Breyman, "Knowledge as Power: Ecology Movements and Global Environmental Problems," in *The State and Social Power in Global Environmental Politics*, ed. Ronnie Lipschutz and Ken Conca (New York: Columbia University Press, 1993); Sheila Jasanoff, "NGOs and the Environment: From Knowledge to Action," *Third World Quarterly* 18, no. 3 (1997): 579–594.

Attention has also been paid to the role of international environmental NGOs outside of the state-based international environmental treaty process. Paul Wapner, for example, has highlighted the importance of these groups in environmental politics more broadly. They have enormous influence through, for example, their role in the mass media in raising the general public's sensibility to environmental issues. This publicity has a profound impact on peoples' consumption patterns and the demands that they make of their states to act on environmental issues. Environmental NGOs also put pressure directly on corporations to change their environmental behavior.[27] In the case of hazard transfer, environmental NGOs have directed their campaigns not just at states but also at public awareness. They have also attempted to influence industry players directly through campaigns planned to embarrass specific corporations.

Each of these dimensions of the nonstate actor literature is extremely important to our understanding of the role of environmental NGOs in global politics and in the case of hazard transfer in particular. What the studies on nonstate actors are lacking, though, and what the hazard transfer case highlights, is the efforts of other nonstate actors, specifically TNCs and business lobby groups, to influence the process. This situation is only now changing. Growing interest in these players can partly be explained by their increased visibility at global environmental negotiations since UNCED, where they were encouraged by the organizers of that conference to participate openly in the international dialogue on environment and development.[28]

Focus on these actors is not entirely new. Global firms have, to varying degrees, been on the research agenda in the field of international political economy for the past 30 years. The expanding reach of global corporations in the 1970s led many to study their implications for the authority and power of states.[29] A strong resurgence of interest in these actors has occurred, as their numbers and economic power have grown phenomenally over the past decade, raising anew concerns about state capability to

---

[27] Paul Wapner, *Environmental Activism and World Civic Politics* (Albany: SUNY Press, 1996).

[28] Chatterjee and Finger 1994; Harris Gleckman, "Transnational Corporations' Strategic Responses to 'Sustainable Development,'" *Green Globe Yearbook* (Oxford: Oxford University Press, 1995); Matthias Finger and James Kilcoyne, "Why Transnational Corporations Are Organizing to 'Save the Global Environment,'" *Ecologist* 27, no. 4 (1997): 138–42.

[29] Robert Gilpin, *US Power and the Multinational Corporation* (New York: Basic Books, 1975); Raymond Vernon, *Sovereignty at Bay* (New York: Basic Books, 1971).

control their activities.³⁰ The influence of TNCs in the global political economy has reached unprecedented levels. There now exist some 38,000 TNCs with a total of over 250,000 affiliates worldwide.³¹ Production by these global firms is now worth more than global trade, while foreign direct investment stock in the early 1990s grew at twice the pace of trade.³² Recognizing the power that arises from this presence, many scholars are asserting the need for further research on TNCs and their role in global politics.³³

Although most studies on TNCs in the field of international relations in the 1970s tended to ignore their impact on the environment, there is now rising concern over the role they play in this respect.³⁴ Indeed, growing attention has been paid to their direct contribution to global environmental problems. Industries with a high degree of environmental impact such as natural resource extraction, chemicals, and electronics tend to be dominated by TNCs. The global nature of TNCs has enabled them to access supplies around the world and to take advantage of cost differentials. Some authors claim that their profit-oriented focus and their global nature make it easy for them to ignore the environmental impacts of their business.³⁵ Others have argued that the situation is much more complex. Peter Dauvergne, for example, has argued that the combination of domestic political situations and the structure of Asian transnational timber companies makes it nearly impossible for them to act in an environmentally oriented fashion.³⁶

But global corporations are not the only ones whose economic activities have environmental consequences that influence global environmental politics. Industry lobby groups have grown in numbers in a way that mirrors the growth in environmental NGOs. By the late 1990s the pres-

---

³⁰ John Stopford and Susan Strange, *Rival States, Rival Firms* (Cambridge: Cambridge University Press, 1991); Mark Zacher, "The Decaying Pillars of the Westphalian Temple: Implications for International Order and Governance," in *Governance without Government*, ed. James Rosenau and Ernst-Otto Czempiel (Cambridge: Cambridge University Press, 1992).
³¹ UNCTAD Division on TNCs and Investment, *World Investment Report, 1995: Transnational Corporations and Competitiveness* (New York: United Nations, 1995), 9.
³² Ibid., 3.
³³ James Rosenau, "Governance in the Twenty-first Century," *Global Governance* 1, no. 1 (1995): 13–43; Stopford and Strange 1991.
³⁴ Nazli Choucri, "Multinational Corporations and the Global Environment," in *Global Accord*, ed. Choucri (Cambridge, Mass.: MIT Press, 1993).
³⁵ Karliner 1997; Korten 1995.
³⁶ Peter Dauvergne, *Shadows in the Forest* (Cambridge, Mass.: MIT Press, 1997); see also Dauvergne, "Corporate Power in the Forests of the Solomon Islands," *Pacific Affairs* 71, no. 4 (1998–99): 524–46.

ence of industry lobby groups at the negotiation of global environmental treaties had become routine, particularly negotiations with clear implications for industry. Some of the more prominent examples are negotiations regarding the waste trade, climate change, ozone depletion, biodiversity, and deforestation. Business actors traditionally have tried to influence global environmental matters by lobbying the state at the domestic level. Industry groups and corporations, for example, tend to influence government positions in global environmental negotiations, as occurred in the United States during the discussions leading up to the negotiation of the Montreal Protocol.[37] While this is an important aspect of industry's efforts to ensure that treaties which states enter into are in accordance with industry's desires, business actors are increasingly focusing their lobbying efforts directly at the international level. This shift is in part a response to a steadily increasing number of extranational environmental regulations. Perhaps more important, it is also a reaction to the rise in environmental NGO activity at this level. So while business may prefer to deal with national governments as the focus of its lobbying efforts because that terrain is more familiar to business interests, it is finding itself in the position of having to lobby at the global level as well, alongside environmental NGOs.[38]

The analysis of industry's influence in global environmental governance has thus far mainly been directed at its role in persuading states to adopt positions that secure industry's economic interests. The power these players have by virtue of their weight in states' economies is seen by some to be key in explaining the influence that industry does indeed seem to have in global negotiations.[39] There has also been a growth in research on industry's attempts to self-regulate at the global level, through the adoption of voluntary codes of environmental conduct.[40] In putting the role of business actors at the center of the study of global environmental politics, this

---

[37] David Levy, "Business and International Environmental Treaties: Ozone Depletion and Climate Change," *California Management Review* 39, no. 3 (1997): 54–71.
[38] This point is made by Levy and Egan in the case of the climate change regime. David Levy and Daniel Egan, "Capital Contests: National and Transnational Channels of Corporate Influence on the Climate Change Negotiations," *Politics and Society* 26, no. 3 (1998): 343.
[39] Peter Newell and Matthew Paterson, "A Climate for Business: Global Warming, the State and Capital," *Review of International Political Economy* 5, no. 4 (1998): 679–703; Levy and Egan 1998: 337–61.
[40] Riva Krut and Harris Gleckman, *ISO 14001: A Missed Opportunity for Sustainable Global Industrial Development* (London: Earthscan, 1998); Jennifer Clapp, "The Privatization of Global Environmental Governance: ISO 14000 and the Developing World," *Global Governance* 4, no. 3 (1998): 295–316.

research helps to explain outcomes and confirms the weakness of a state-centric regime-based approach to studying global environmental issues.

The hazard transfer issue highlights some key trends with respect to TNCs in global environmental politics. First, globally connected firms were the ones exporting toxic wastes in the first place, making the shift toward hazardous waste exports for recycling as well as transferring hazardous production processes and outmoded toxic equipment. They are thus contributing directly to the problem of hazard relocation and are the ones toward which regulations are ultimately directed. Yet they have been able to use the global economy to evade regulations by finding alternative channels for hazard transfer in the face of more stringent regulations. Second, business lobby groups have taken a key role alongside NGOs in the negotiation of environmental agreements regarding the waste trade. Not only are they involved in the more public diplomatic meetings, but they also aim to influence outcomes via less public technical meetings. The role of industry in more technical and less public aspects of global environmental politics, such as interpreting and implementing agreements, has received somewhat less attention in the literature. Yet this is an area in which these actors appear to have a great deal of influence. In certain issue areas where the regulation of industrial activity is at the center of international policy, such as in the case of the waste trade, industry's input into interpretation and implementation has been significant. Third, the hazard transfer issue highlights a trend toward the privatization of global environmental governance through the development of voluntary environmental management standards for industry. Environmental NGOs were excluded from the development of ISO 14000 standards and the majority of developing countries participated only marginally in the process.

Map of This Book

The second chapter of this book gives the history of the rise of the trade in hazardous wastes from rich to poor countries and the politics of the negotiation of the Basel Convention as the major international treaty seeking to address the problem. I argue that the emergence of the waste trade problem is intricately linked to global economic factors. I also argue that environmental NGOs carved out a significant role for themselves early on in the treaty negotiation process. In Chapter 3 I outline the way in which the weaknesses of the initial Basel Convention fostered new waste trade problems, such as the growth in the trade in toxic wastes for recycling purposes. I also trace the growing pressure by environmental NGOs and developing countries to amend the convention to address those weaknesses. Here I argue that environmental NGOs had an instrumental role in achieving this outcome, in particular through their strong alliance with developing country states and their effective campaign strategies. In Chapter 4 I examine the growing participation of industry lobby groups in the Basel process. I argue that although these groups entered somewhat late into the political process, they have been able to exert considerable influence over Basel Convention politics through their effective lobbying on more technical issues such as the definition of what wastes are to be classified as hazardous under the convention. They also continue to wage battle to reverse the Basel Ban Amendment. I also outline the changes in the environmental NGO strategies after the Basel ban was adopted and their response to the growing importance of industry lobby groups in the process.

In Chapter 5, I look at another potential outcome of the increased regulation on the trade in hazardous wastes, that of the migration of hazardous industries from rich to poor countries. I argue that the mainstream literature on this topic has tended to underestimate the forces that push toward the relocation of hazardous industries to developing countries, particularly after the adoption of the Basel Convention. In Chapter 6 I examine the prospects for market-based and voluntary initiatives to promote cleaner production on a global scale and the transfer of cleaner production technologies to developing countries in particular. I argue that the growth of investment in the "environment industry" in developing countries has thus far focused more on cleaning up toxic messes than on preventing them in the first place. I also argue that discussion of a set of global-level performance standards for industry has been preempted by industry's embrace of voluntary environmental management system measures such as the ISO 14000 standards. Yet these latter standards have tended to reinforce, rather than reduce, the gap in regulations on hazardous waste and production processes between rich and poor countries. For this reason they are unlikely to lead to the types of changes necessary to install truly clean production in the developing world.

In the concluding chapter I discuss recent trends that illustrate the ongoing evolution of the hazard transfer problem, including the growing problem of hazard transfer between poorer countries. I argue that the most promising way to address the hazard transfer problem will be to avoid the generation of hazardous wastes in the first place. Toward this end I outline a strategy for the promotion of clean production on a global scale. I argue that a globally agreed framework that incorporates a number of features simultaneously will likely be the only way to achieve this goal.

These measures include commitment to strengthening the Basel Convention; active promotion of the concept of clean production in government regulations; global requirements for transnational corporations in hazardous industries with respect to environmental performance and information disclosure; and continued pressure from environmental NGOs to help monitor these efforts. Though many of these measures may be politically difficult to implement with the force necessary, they may be the only way to stop the dynamics of the global hazards problem in the context of a world on disparate economic planes.

The displacement of hazardous wastes and the technologies that generate those wastes from richer to poorer countries has been a disturbing feature of the global political economy. It is disturbing especially because its incidence emerged at the same time that concern for global environmental problems was growing. Since the mid-1980s an unprecedented number of international environmental agreements have been made. But the fact that agreements are being negotiated and signed and even implemented does not necessarily mean that the original problem that they seek to remedy is addressed satisfactorily. The case of hazard transfer from rich to poor countries illustrates this point very clearly. It seems that every 'victory' by the groups trying to halt the trade is tempered by yet another outlet for hazards that takes on more importance for both those trying to stop it, and those trying to maintain the opening. At each turn we are left wondering if we are on a path toward greener, cleaner production, or toward continued hazard migration in a new form.

2
The Hazardous Waste Trade and International Regulatory Measures

The 1980s were marked by an increasingly global economy, facilitated by the growing political thrust toward trade and financial liberalization in both rich and poor countries. This situation gave hazardous waste generators and handlers the ability to respond at a global level to factors that affected their costs of disposal and treatment. The weak position of developing countries in the global political economy made these countries especially vulnerable to waste exports from richer countries. Imports of wastes into these countries grew in this period, and they were extremely difficult to control. The political response to the growing cross-border trade in hazardous wastes was to put in place international regulations. The attempt to deal with the problem primarily at the national level was not sufficient to counteract a problem that was by nature transboundary.

As a recent report of the OECD has noted, "Hazardous waste management has become increasingly a globalized business, requiring global regulatory systems in light of the potential environmental effects of improper practices."[1] From early on in the process to develop an international set of rules to control the trade in toxic wastes, it was obvious that states were not the only actors who had a stake in the issue. Environmental NGOs, as well as industry groups involved in the waste trade, also were deeply involved in the negotiation process. The result, the Basel Convention on the Transboundary Movement of Hazardous Wastes and Their Disposal, as well as regional waste trade agreements, were the products not just of state bargaining but also of significant involvement by nonstate actors.

This chapter outlines the rise of the waste trade and the international political response to it. I argue that global economic factors played an important part in the rise of the waste trade. I then assert that the importance of nonstate actors in the political response to the waste trade can be partly explained by the global economic nature of the waste trade problem itself, which at the time it emerged was largely outside of the control of states. States were a large part of the focus of environmental groups for action on international rules. But these groups also launched a much broader campaign. They attempted to influence waste traders directly by embarrassing them through a public awareness campaign. This effort required a significant amount of research on their part, which in turn raised their credibility among states by giving them unparalleled expertise on the waste trade.

Toxic Waste and the Global Economy

The export of toxic waste to less industrialized countries can be best explained in the context of the current global economy. The cross-border trade in toxic waste began as a general practice in the late 1970s and continued to grow throughout the following decades. Just a handful of industrialized countries produce 95 percent of the world's hazardous waste.[2]

Most accounts of the waste trade focus on the push factors spurred by environmental concerns in industrialized countries. Following incidents

---

[1] OECD, *Trade Measures in the Basel Convention on the Control of Transboundary Movements of Hazardous Wastes and Their Disposal*, COM/ENV/TD(97)41/FINAL (Paris: OECD, 1998), 7.
[2] The per annum global generation of hazardous waste totals some 400 million tons per year. The United States generates 85 percent of the world's hazardous wastes, and the European Union countries generate 5–7 percent of the world total. "Basel Convention—More Action?" *Environmental Policy and Law* 23, no. 1 (1993): 14.

of mishandled hazardous wastes such as occurred at Love Canal in the United States and Seveso in Italy in the 1970s and early 1980s, the regulations accompanying hazardous waste disposal and the costs associated with it grew tremendously. As a manifestation of the NIMBY (not in my backyard) syndrome, people in industrialized countries were increasingly unwilling to have the dump sites located nearby unless they were strictly regulated to ensure that such incidents could not occur. The result was that regulations in hazardous waste–producing countries became very stringent. Landfill capacity for toxic wastes began to decline, and resistance to new landfills became more vocal.[3] As a direct consequence of these developments, costs of local hazardous waste disposal in these countries soared. For example, in the United States landfill costs for dumping hazardous waste rose from U.S.$15 per ton in 1980 to U.S.$250 per ton in 1988.[4]

In addition to the push factors, there were also pull factors. Lower waste disposal fees in countries with less stringent environmental regulations encouraged the movement of waste across borders from rich to poor countries. The owners of dump sites in less industrialized countries generally charge much lower fees for disposal of imported waste than do those in more industrialized countries. The weak financial position of developing countries in the global political economy helps to explain why these countries had especially low waste disposal costs. The poorest and weakest countries burdened with international debts were desperate for foreign exchange and often were receptive to any proposal that would enable them to earn hard currency. Although the amount paid to recipients in the developing world was large compared to their need for foreign exchange, it was small compared to costs for disposal in advanced industrialized countries. In the mid-1980s, the disposal cost per metric ton of hazardous waste in Africa was around U.S.$40, and in some cases as low as U.S.$2.50 which was significantly lower than costs in the industrialized countries at the time.[5] These costs were low because toxic waste disposal facilities and regulations governing toxic waste disposal were virtually nonexistent in developing countries. Most of these countries did not produce such waste themselves and therefore lacked expertise on its proper treatment. With the financial incentives for accepting it in place and little knowledge of its effects, the developing world did little to develop stringent waste disposal

---

[3] OECD 1998, 7.
[4] Laura Strohm, "The Environmental Politics of the International Waste Trade," *Journal of Environment and Development* 2, no. 2 (1993): 133.
[5] Mostafa Tolba, "The Global Agenda and the Hazardous Wastes Challenge," *Marine Policy* 14, no. 3 (1990): 205–206.

regulations. With the onset of a serious economic crisis among these countries in the 1980s, toxic waste soon made its way to nearly every corner of the developing world.

These push and pull factors might not have been so important had it not been for an increasingly global and fluid international trading system. Extensive transport and communications networks put in place for trade in commodities have facilitated the trade in toxic wastes. Lower transportation and communication costs, the relative ease with which trade routes are established and abandoned, and the difficulties in checking every import container, especially in developing countries, have encouraged the export of hazardous wastes. This fluidity of global trade has been enhanced as developed and developing countries alike have adopted more liberal trade policies over the past twenty years. These qualities of global trade today have made the export of hazardous waste to poorer countries a simple and lucrative business for waste trade entrepreneurs. Though they are underplayed in many accounts of the waste trade, the global economic factors have been extremely important in explaining the rise of the trade.

It is extremely difficult to quantify precisely the global generation of hazardous wastes and the extent of the international waste trade. Most of the export deals have been clandestine, and therefore difficult to track. In addition, no two countries have the same definition of exactly what constitutes a hazardous waste. As a result, recorded waste trade transactions do not follow a uniform standard. Despite these difficulties in measurement, the United Nations Environment Programme estimates that around 440 million metric tons of toxic waste were produced every year in the 1990s, up from around 300–400 million metric tons per year in the 1980s. Most of the world's toxic waste is generated in the United States and Western Europe.[6] Table 2.1 gives an indication of the amounts of hazardous waste generated in various OECD countries. These hazardous wastes generally include substances such as PCBs, dioxins, asbestos, heavy metals, and certain plastics. Box 2.1 outlines some of the waste streams and compounds, along with accompanying health effects.

Roughly 10 percent of all toxic waste generated globally is estimated to make its way across international borders.[7] Although this may seem to be a small proportion of all hazardous wastes generated, it is still a significant

---

[6] Preface to 1999 version of the Basel Convention.
[7] Christoph Hilz, *The International Toxic Waste Trade* (New York: VanNostrand Reinhold, 1992), 20. There are huge gaps in figures from the OECD. Recent data indicate that about 2 million tons of toxic waste were exported by the OECD countries in 1989 and 1990, but not all member countries reported. See *ENDS Report*, no. 223 (August 1993), 15.

Table 2.1 Generation of Hazardous Wastes in Select OECD Countries

Country[a]	Year	(1,000 metric tons)
Australia	1992	426
Austria	1995	915
Belgium[b]	1994	27,530
Canada	1991	5,896
Denmark*	1993	91
Finland	1992	367
France	1992	7,000
Germany	1993	9,020
Greece*	1992	450
Ireland*	n.d.	66
Italy*	1991	3,387
Japan	n.d.	666
Netherlands[c]	1993	2,600
New Zealand	1990	110
Norway	1991	220
Portugal	1994	1,365
Spain*	1987	1,708
Sweden	1985	500
Switzerland*	1993	837
Turkey*	1989	300
United Kingdom*	1993	1,957
United States[d]	1993	258,000

Source: OECD, *Transfrontier Movements of Hazardous Wastes, 1992–93 Statistics* (Paris: OECD, 1997).
n.d.: no data provided
[a] Most of these data have been communicated directly to the Secretariat of the Waste Management Policy Group with the exception of a few countries (marked with an asterisk) for which the source of the data is the Compendium 1995 of OECD Environmental Data.
[b] The figure for the generation of hazardous wastes in Belgium includes all wastes produced in the Wallonia region only by the industry sector such as residues from mining operations (about 3 MT), glass, wood, paper, food, and ferrous and nonferrous metals.
[c] Netherlands hazardous wastes generation includes 845,000 metric tons of contaminated soil.
[d] The difference between the waste generation figures for the United States and Europe arises largely because the United States defines large quantities of dilute dishwater as hazardous wastes while in Europe, these materials are managed under water protection regulations.

Box 2.1 The Health Impacts of Toxic Waste: Some Examples

Commonly Exported Waste and Waste Components	Health Impact
Polychlorinated Biphenols (PCBs)	Reproductive dysfunction, immune system suppression
Asbestos	Lung disease, cancer, can lead to abestosis causing disability or death
Chlorine	Respiratory problems, skin irritant
Dioxins	Reproductive disorders, immune system suppression
Chromium	Liver and respiratory problems, causes allergic responses to skin
Chlorinated Solvents	Neurological damage, liver problems
Banned Pesticides	Most are carcinogenic, highly poisonous
Incinerator Ash: often contains metals such as lead, mercury, arsenic, chromium	Lead: Neurological disorders, kidney and nervous system damage Mercury: Neurological, liver, and kidney damage, coma and death, dangerous during pregnancy Arsenic: Lung and tissue damage, liver and kidney injury, potentially fatal
Lead-acid batteries	Contains lead
Metal waste and scrap: often contains lead, mercury, copper, cadmium	Copper: Liver damage Cadmium: Kidney damage, respiratory problems, cancer, irritates digestive tract
Plastic wastes: often contains polyvinyl chloride (PVC)	Packaging waste: danger from toxins or bacteria from residues of previous contents, releases highly toxic fumes when recycled or incinerated PVC: liver and nerve damage, cancer. Generally contains additives such as toxic metals like cadmium, lead, phthalate plasticizer
Scrap tires	Long-lasting respiratory difficulties. Shredding or recycling can lead to exposure to carcinogenic hydrocarbons, incineration releases highly toxic fumes.

Computer and electronic scrap: contains PVC, heavy metals, and other materials such as PCBs	Releases highly toxic fumes when recycled
Cable scrap: often sheathed in lead, copper, or PVC	Immune system suppression, reproductive dysfunction, contains human carcinogens
Sewer sludge: often contain mercury, lead, cadmium, PCBs and dioxins	Can be a transporter for parasites and human disease Can interfere with body hormone balances
Furnace dust: electric arc furnace steel dust contains zinc, lead, nickel, and dioxins; copper smelter furnace dust contains lead and cadmium	Respiratory problems, lung cancer from dust inhalation

amount. For example, a shipment of toxic waste is estimated to cross a border within the Organization for Economic Cooperation and Development states alone every five minutes, twenty-four hours a day, 365 days a year.[8] The bulk of the waste trade is conducted among industrialized countries themselves. Table 2.2 shows the amounts of hazardous waste traded by OECD countries from 1989 to 1993. Estimates of the amount of the trade in toxic waste to countries outside of the OECD vary. Some have estimated in the early 1990s that around 20 percent was with non-OECD countries, including some 10–15 percent to Eastern Europe and the rest to developing countries.[9] But others have estimated that over half of the trade in toxic wastes at that time was with non-OECD countries, with about 20 percent going to developing countries.[10] Though these estimates represent a wide range, they signal that a significant proportion of waste exports has made its way to countries that lack the economic resources and regulations to ensure that they are disposed of properly. Table 2.3 gives an indication of the number and fate of these waste movements from OECD to non-OECD countries. It is important to note that some have downplayed the magnitude of the waste trade with developing countries in the late 1980s and early 1990s, pointing out that many of the proposed shipments were turned back and not completed. For this reason some have

[8] William Long, "Economic Aspects of Transport and Disposal of Hazardous Wastes," *Marine Policy* 14, no. 3 (1990): 199.
[9] Jonathan Krueger, *International Trade and the Basel Convention* (London: RIIA, 1999), 14.
[10] Hilz 1992, 20–21; Christoph Hilz and Mark Radka, "Environmental Negotiations and Policy: The Basel Convention on Transboundary Movement of Hazardous Wastes and Their Disposal," *International Journal of Environment and Pollution* 1, no. 1–2 (1991): 56.

Table 2.2 Summary of Transfrontier Movements of Hazardous Wastes from 1989 to 1993 in OECD Countries (in metric tons)

	Exports					Imports				
	1989	1990	1991	1992	1993	1989	1990	1991	1992	1993
Australia[a]	500	1,000	3,200	275	0	0	0	0	0	0
Austria[b]	86,773	68,162	82,129	70,023	83,998	50,981	19,180	111,595	79,107	28,330
Belgium[c]	176,983	491,784	645,636	37,278	34,073	1,036,260	1,070,496	1,021,798	208,052	236,010
Canada[d]	101,083	137,818	223,079	174,682	229,648	150,000	143,811	135,161	123,998	173,416
Denmark	8,120	9,214	21,758	15,858	n.d	11,401	16,376	15,200	100,244	n.d.
Finland	64,665	19,174	24,174	21,757	20,628	7,565	9,889	4,605	5,145	4,770
France	n.d.	10,552	21,126	32,309	78,935	n.d.	458,128	636,647	512,150	324,538
Germany[e,l]	990,933	522,063	396,607	548,355	433,744	45,312	62,636	141,660	76,375	78,219
Greece	n.d.	305	n.d.	n.d.	n.d.	n.d.	n.d.	n.d.	n.d.	n.d.
Iceland	n.d.	90	151	n.d.	n.d.	0	0	0	n.d.	n.d.
Ireland	13,808	n.d.	n.d.	n.d.	n.d.	n.d.	n.d.	n.d.	n.d.	n.d.
Italy[f]	10,800	19,968	13,018	21,627	19,365	0	0	0	n.d.	n.d.
Japan	40	0	n.d.	n.d.	n.d.	5,125	397	n.d.	n.d.	n.d.
Luxembourg	n.d.	n.d.	n.d.	n.d.	n.d.	n.d.	n.d.	n.d.	n.d.	n.d.
Netherlands[g]	188,250	195,377	189,707	172,906	163,180	88,400	199,015	107,251	250,355	236,673
New Zealand[h]	200	0	21	208	10,469	0	0	0	0	n.d.
Norway[i]	8,078	16,535	14,636	14,545	16,639	0	0	2,415	640,701	81,207
Portugal[j]	n.d.	1,954	292	457	815	n.d.	0	1,147	5,638	7,195
Spain[k]	280	20,213	6,578	15,803	13,943	27,413	82,269	81,597	66,356	104,716
Sweden	45,015	42,636	63,801	22,185	22,484	33,863	47,223	34,195	61,725	82,933
Switzerland	108,345	121,420	126,564	132,138	125,840	7,684	6,688	6,416	10,471	8,360
Turkey	0	0	n.d.	n.d.	n.d.	0	0	n.d.	n.d.	n.d.

United Kingdom[l]	0	496	857	0	40,740	34,983	54,074	44,673	66,294
United States[m]	118,927	118,416	108,466	145,556	142,709	n.d.	n.d.	n.d.	n.d.	n.d.

Source: OECD, *Transfrontier Movements of Hazardous Wastes, 1992–93 Statistics* (Paris: OECD, 1997)

n.d.: no data provided

Because of differences in national definitions of hazardous wastes, great caution should be exercised when using these figures.

[a] Australian data refer to fiscal year (from July 1 to June 30) and concern permits for final disposal.
[b] Austria enforced its new ordinance on hazardous wastes in 1991.
[c] Belgium data include toxic wastes as well as household refuse, and recyclable nonferrous metals.
[d] Canada enforced its new legislation on transfrontier movements of hazardous wastes in November 1992.
[e] Differences between 1989 and 1990 data are largely due to German unification in 1990.
[f] Export data for 1989 are an estimate based on figures available for the last three months of the year.
[g] Dutch data excludes imports and exports of nonferrous metals waste destined for recycling.
[h] Until 1992: PCBs exports only. In 1993: exports of hazardous wastes going to recovery only.
[i] The increase of imports in 1992 is due to huge amounts of aluminium salt slag being now sent to Norway to be recovered.
[j] Portugal enforced its new legislation on transfrontier movements of hazardous wastes in 1992.
[k] Spain changed its regulations concerning hazardous wastes between 1989 and 1990.
[l] Only wastes going to final disposal have to be notified.
[m] Until new legislation is passed to implement the Basel Convention, the United States requires written notice and consent for exports only.

Table 2.3 Number of OECD to Non-OECD Waste Trade Schemes by Year

Year	Further use claimed	Final disposal	Total where fate/pretext is known	Fate/pretext unknown	Total of all schemes	Percentage of "further use" known schemes
1989	54	17	71	5	76	76%
1990	92	19	111	7	118	83%
1991	94	14	108	5	113	87%
1992	238	30	268	17	285	88%
1993	123	15	138	8	146	89%
Totals	601	95	696	42	738	86%

Source: Greenpeace, *Database of Known Hazardous Waste Exports from OECD to Non-OECD Countries, 1989–March 1994* (Amsterdam: Greenpeace International, 1994).

Note: For the purpose of this table, schemes listed as having taken place in multiple years are listed for each year. This explains why the totals are greater than those found in the other tables. Also this table does not include schemes for 1994.

This table lists schemes that have been claimed to be destined for recycling or some form of "further use," those claimed for final disposal, and those that are unknown. It is significant to note the percentage of those schemes claimed for recycling or further use of the total where the fate is known.

questioned the extent to which the problem was a "crisis" at that time.[11] However, the fact that there were so many attempts to ship the waste to poor countries, even if not all of them were carried out, indicates there were strong incentives to transfer hazards via this method.

The serious discrepancies in the data on the waste trade between rich and poor countries are somewhat understandable given the nature of this trade. Much of the toxic waste trade with developing countries has taken place along what many consider to be dark and illegal channels of the global economy. Most generators of waste contract out disposal of their toxic by-products to waste dealers. Some dealers involved in waste shipments between rich and poor countries in the 1980s worked through umbrella companies based in tax havens with bogus company directors. Once a transaction was complete, the firm in question was often dissolved, and the waste traders simply created another for the next waste deal. These fly-by-night operations made tracing the waste to the dealer, much less to the generator, nearly impossible. Waste dealers from industrialized countries who sought to dispose of wastes in non-OECD countries were careful to keep their operations somewhat covert. If the true nature of the export was discovered, they might be held liable for damages and be responsible for the waste's removal. Such firms sought anonymity and protected one another in maintaining their secrecy.[12] Waste import deals were for the most part contracted by individual entrepreneurs in recipient countries who have also sought to maintain anonymity.

Despite the difficulties in tracking all of the trade that has taken place, there are hundreds of documented cases of toxic waste exports and attempted exports from rich to poor countries in the late 1980s to the early 1990s. This traffic in hazardous wastes was a major reason behind the huge push to negotiate international rules to govern it. A number of cases have highlighted the serious problems associated with this trade. As these cases came to light, concerns were raised. Many argued that it was unjust to send hazardous wastes to countries that had nothing to do with their generation, and that did not receive direct benefits from the goods produced which resulted in these wastes, yet had to bear the environmental and health consequences from those wastes in exchange for much needed foreign exchange. It was on these grounds that environmental groups and

---

[11] Mark Montgomery, "Reassessing the Waste Trade Crisis: What Do We Really Know?" *Journal of Environment and Development* 4, no.1 (1995): 1–28.
[12] François Roelants du Vivier, "Control of Waste Exports to the Third World," *Marine Policy* 14, no. 3 (1990): 5.

developing countries labeled the waste trade with developing countries as "toxic colonialism," and "toxic terrorism."

Developing Country Targets for Waste

Africa and other less industrialized countries became favorite targets for waste traders in the mid-1980s, as shown in Tables 2.4 and 2.5. These countries generally had weak environmental laws and very limited state control over customs officials who approved import shipments. The marginal position of Africa in particular in the global economy encouraged waste exports to the continent. African countries, many gripped by poverty, war, and famine, were in desperate need of the foreign exchange to be gained from offering dump sites in the 1980s. Over half of the countries on that continent had been approached to accept hazardous wastes by 1990.[13] Other regions also targeted for waste exports in the late 1980s were the South Pacific, the Caribbean, and Latin America. When opposition to the waste trade emerged from these regions by the early 1990s, Asia and Eastern and Central Europe were increasingly targeted by waste traders.[14] According to data compiled by the environmental group Greenpeace, between 1989 and early 1994, there were 299 known attempted or completed toxic waste dumping incidents in Eastern and Central Europe and 239 such incidents in Asia, compared with 148 incidents in Latin America and the Caribbean, 30 in Africa, and 12 in the Pacific for that same period.[15]

Not all of the attempts to export waste to non-OECD countries were completed. But of those that were, the conditions under which the wastes were dumped were more often than not far from ideal. The industrialized countries had enough problems making their own landfills safe, and most dump sites in the developing world were not even regulated.[16] Much of the

---

[13] On the African case, see Charles Anyinam, "Transboundary Movements of Hazardous Wastes: The Case of Toxic Waste Dumping in Africa," *International Journal of Health Services* 21, no. 4 (1991); Mutombo Mpanya, "The Dumping of Toxic Waste in African Countries: A Case of Poverty and Racism," in *Race and the Incidence of Environmental Hazards*, ed. Bunyan Bryant and Paul Mohai (Boulder: Westview, 1992); Jennifer Clapp, "Africa, NGOs, and the International Toxic Waste Trade," *Journal of Environment and Development* 3, no. 2 (1994): 17–46.
[14] See Jim Puckett, "Disposing of the Waste Trade: Closing the Recycling Loophole" *Ecologist* 24, no. 2 (1994): 54.
[15] Greenpeace International, *Database of Known Hazardous Waste Exports from OECD to Non-OECD Countries, 1989–March 1994* (Amsterdam: Greenpeace International, 1994).
[16] On the problems and risks in toxic waste production and disposal, see, for example, Robert Allen, *Waste Not, Want Not: The Production and Dumping of Toxic Waste* (London: Earthscan, 1992); Brian Wynne, *Risk Management and Hazardous Waste* (Berlin: Springer-Verlag, 1987).

Table 2.4 Number of Schemes Proposed for Exports by Receiving Region and Year

Region	1989	1990	1991	1992	1993	Totals
Baltics and Eastern/Central Europe	32	50	43	113	61	299
Africa	11	4	4	7	4	30
Pacific	1	4	1	2	4	12
East Asia	4	14	22	50	22	112
Southeast Asia	0	2	10	46	26	84
South Asia	2	3	2	24	12	43
Middle East	0	0	1	12	1	14
Latin America/Caribbean	27	43	30	32	16	148
Totals	77	120	113	286	146	742

Source: Greenpeace, *Database of Known Hazardous Waste Exports from OECD to non-OECD Countries, 1989–March 1994* (Amsterdam: Greenpeace International, 1994).

Note: For the purposes of this table, schemes listed as having taken place in multiple years are listed for each year. This explains why the totals are greater than those found in other tables. Also this table does not include schemes for 1994.

waste that was exported to less industrialized countries was dumped either in flimsy containers or in no containers at all. In the hot and wet climates of tropical countries, the waste could easily leach into the soil and the water table. Wastes shipped to Eastern and Central Europe have also been poorly contained. In some cases wastes were shipped in corroded barrels that subsequently ruptured under extreme weather conditions.[17]

One of the first and most notorious examples of waste traders seeking to offload their toxic cargo is the voyage of the ship the *Khian Sea*. This ship set sail from the United States in 1986 loaded with nearly fourteen thousand tons of toxic fly-ash from Philadelphia's municipal waste incinerator. After an unsuccessful attempt to dump the ash in the Bahamas, the ship sailed the Caribbean Sea in search of a port that would accept the waste. After months of searching, the ship was authorized to unload the cargo in Haiti, under the label of fertilizer. When the government became aware of

---

[17] See, for example, Andreas Bernstorff and Katherine Totten, *Romania: Toxic Assault* (Hamburg: Greenpeace Germany, 1992); Andreas Bernstorff and Jim Puckett, *Poland: The Waste Invasion* (Amsterdam: Greenpeace International, 1992); A. Bernstorff et al., *Russia: The Making of a Waste Colony* (Moscow: Greenpeace Russia, 1993).

Table 2.5 Results of Hazardous Waste Trade Proposals from OECD to Non-OECD Countries 1989–93

Status	1989	1990	1991	1992	1993	Total
Actual	5	16	30	155	72	278
Rejected	31	41	28	25	10	135
Stopped/returned	7	18	7	27	16	75
Proposed/planned	1	3	4	13	14	35
Other/unknown/abandoned	25	38	39	48	14	144
Total	69	106	98	268	126	

Source: Jonathan Krueger, "Prior Informed Consent and the Basel Convention: The Hazards of What Isn't Known," *Journal of Environment and Development* vol. 7, no.2 (1998), p. 126. Krueger's calculations are based on *Greenpeace, Greenpeace Database of Known Hazardous Waste Exports from OECD to non-OECD Countries, 1989–March 1994* (Amsterdam: Greenpeace International, 1994).
- Number of total known waste export schemes: 667.
- Number of shipments resulting in trade for disposal or recycling: 278 (41.7%).
- Number of shipments rejected by importing state: 135 (11.2%).
- Number of shipments stopped by exporting state or returned to exporting state: 75 (11.2%).
- Number of shipments proposed/planned (without final result): 35 (5.2%).
- Number of abandoned shipments or schemes with unknown/other results: 144 (21.6%).

the scheme, it ordered the waste to be removed. The ship left Haiti in search of an unsuspecting country to take the load but left behind an estimated four thousand tons of the ash on the beach in Haiti. The ship's crew tried in vain to unload the remainder of toxic ash in Africa, Europe, the Middle East, and East Asia. The wide publicity of the voyage by environmental groups and the media ensured that no government would accept it. After twenty-seven months of trying to find a dump site, the ash mysteriously disappeared from the ship in Southeast Asia. Many believed it was dumped at sea.[18] The remainder of the waste continued to sit on the beach in Haiti where it was originally dumped. In 1998 efforts began to have the waste returned to the United States for proper disposal. It was not removed until April 2000, and as of early 2001 it still remained on a barge off the coast of Florida.[19]

---

[18] See Jim Vallette and Heather Spalding, ed., *The International Trade in Wastes: A Greenpeace Inventory* (Washington, D.C.: Greenpeace, 1990): 21–25.

[19] Ramona Smith, "New Ship Hauls Haitian Ash," *Philadelphia Daily News*, October 30, 1998; "Haiti: No Welcome Mat for Return of U.S. Toxic Waste," *Inter Press Service*, June 13, 1999; Victor Fiorillo and Liz Spiol, "Ashes to Ashes, Dust to Dust," *Philadelphia Weekly*, January 18, 2001. Available at http://www.ban.org.

Another of the early major scandals of waste dumping in the developing world involved an incident at Koko, Nigeria, in 1988. A local farmer, Sunday Nana, rented out his backyard for storage to an Italian waste firm. He was paid U.S.$100 per month. The firm subsequently unloaded eight thousand barrels of chemical wastes, brought on the ship *Karin B*. These barrels subsequently burst open after sitting in the hot sun and contaminated the land. The farmer was told that the barrels contained fertilizer. In actual fact, they contained industrial waste contaminated with PCBs and asbestos fibers. Several villagers became extremely ill after they had stolen the barrels and emptied them to use for storage of drinking water. Local residents and Nigerian environmental NGOs were enraged by the incident. The revelation that the barrels contained toxic waste sparked an international scandal, and the Nigerian government eventually forced Italy to take back the wastes.[20]

Other parts of the developing world also received waste exports from rich industrialized countries. Throughout the 1980s Mexico was used as a repository for much of the hazardous waste exported by firms in the United States. Wastes were often sent hidden in other cargo on trucks and trains that crossed the border into Mexico and were subsequently dumped into unregulated landfills.[21] Much of this waste was not detected by border guards, who were more interested in stopping movement of illegal drugs and arms than in preventing toxic wastes from entering Mexico.

Waste traders sometimes took advantage of the weak economic and political situations of certain developing countries. For example, the government of Guinea-Bissau was offered four times the value of its GNP (equal to twice the value of its external debt) if it would accept up to 15 million tons of toxic wastes over a fifteen-year period. The government, which lacked other opportunities to earn hard currency, originally accepted the proposal.[22] After strong pressure from other African governments, however, it officially withdrew from the contract. In 1991, in the midst of famine and war, Somalia received a proposal to accept a waste shipment. This proposal was initially accepted by the health minister of the deposed government, who was reportedly offered a large bribe for

---

[20] For further details, see Bill Moyers and Center for Investigative Reporting (CIR), *Global Dumping Ground* (Cambridge, U.K.: Lutterworth Press, 1991), 1–2; *New African*, no. 253 (October 1988): 22; Economist Intelligence Unit, *Nigeria Country Report*, no. 4 (1988): 8–9.
[21] Moyers and CIR 1991, 41–42; "Enforcement Actions Taken against Polluters on U.S.–Mexico Border," *EPA Environmental News*, June 3, 1992.
[22] Wynne 1989, 121.

letting in the wastes.[23] Although the deal was believed to have been stopped, it has been reported that several European waste trading firms had agreed to pay the Somalis U.S.$80 million to take up to 500,000 metric tons of toxic waste over a period of twenty years. The firms stood to make U.S.$8 to $10 million per shipment.[24] Regarding the case, Mostafa Tolba, at the time executive director of the United Nations Environment Programme, lamented: "Hazardous wastes will always follow the path of lower costs and lower standards."[25]

Some waste exported to the developing world was disguised or labeled as other products, making such shipments even more difficult to track, especially for developing countries. This was the case with the shipment to Zimbabwe in 1984 of over two hundred barrels of "dry cleaning fluid and solvents" which was actually hazardous waste. The waste, sent by a U.S. firm, was discovered by the U.S. Agency for International Development (USAID), which had financed the sale.[26] In another case, fifteen thousand tons of uncontained toxic incinerator ash of U.S. origin was shipped to Guinea in the late 1980s under the label of raw material for building bricks.[27] This waste was dumped on the Guinean island of Kassa, just off the coast of the country's capital, Conakry, by a Norwegian waste management firm. When the pile of ash killed off vegetation and the smell was no longer tolerable to the local residents, an investigation revealed that the material was actually toxic waste. The Norwegian consul general in Guinea was implicated in the incident and was placed under house arrest until the ash was removed. Four government officials were also jailed for their involvement. The waste eventually made its way back to the United States, where it was buried in a landfill.[28] In another case, several U.S. companies attempted to convince the Marshall Islands that imported wastes could be used to build up land mass to ensure the islands would survive possible sea-level rises caused by global warming. The firm that proposed this "land reclamation project" claimed that no hazardous wastes would be involved, but this could not be verified.[29]

---

[23] "Toxic Waste Probe," *West Africa* no. 3917 (October 12–18, 1992): 1735.
[24] "Toxic Waste Adds to Somalia's Woes," *New Scientist* 135 (September 19, 1992): 5; "UNEP Official Urges African Nations to Approve Basel Accord on Waste Shipments," *International Environment Reporter* 15 (October 7, 1992): 654.
[25] "Transfrontier Waste Meeting Focuses on Exports, Liability," *ENDS Report*, no. 215 (December 1992): 37.
[26] Vallette and Spalding 1990, 113; Moyers and CIR 1991, 34–38.
[27] Third World Network, "Toxic Waste Dumping in the Third World," *Race and Class* 30, no. 3 (1989): 46–47.
[28] For further details, see Third World Network 1989, 46–47; Mpanya 1992, 205.
[29] Greenpeace USA, *Pacific Waste Invasion* (Washington, D.C.: Greenpeace, 1992).

In addition to the mislabeling of industrial waste, the export of banned or outdated chemicals and pesticides to developing countries has been a longer-term problem.[30] Some of these chemicals are sold to developing countries despite being banned in the country of export. For example, it has been estimated that the United States alone exported at least fifteen tons of banned pesticides per day in 1991.[31] Many of these chemicals were given to developing countries by industrialized country governments as part of tied-aid packages, in quantities far in excess of the recipients' needs. Backlogs of toxic materials have accumulated because the chemicals have passed their use-by date or have since been banned by the country of import.[32] Thousands of tons of such hazardous chemicals now exist in less industrialized countries and are in need of proper disposal. For example, in the early 1990s the Sudan had substantial stocks of DDT remaining from the 1960s, although the country banned that pesticide in 1980.[33]

The cases of environmentally unsound dumping in poor, less industrialized countries cited above were not isolated incidents. They represent only a small portion of the many diverse schemes waste traders have undertaken or attempted to dispose of unwanted toxic waste. These examples show that the seemingly economic solution of dumping toxic wastes in poor countries, where regulations are few and costs are lowest, is not at all economically or environmentally beneficial for the recipients. The waste compromises economic development prospects in the long run, even if it has brought cash to some countries, or more often to just a few individuals, in the short run. The cleanup costs of hazardous waste dumps were far too expensive for poor countries to meet, as costs in some countries for hazardous waste dump-site cleanup have reached figures in the billions of dollars.[34] The result is that few cleanup efforts were undertaken in the developing world following these incidents.

The subsequent damage to the environment from the import of hazardous waste compromises the economic potential of these developing

---

[30] For an overview, see Barbara Dinham, *The Pesticide Hazard* (London: Zed, 1993), 11–37.
[31] Carl Smith, "U.S. Pesticide Traffic—Exporting Banned and Hazardous Pesticides," *Global Pesticide Campaigner* 3, no. 3 (1993): 1.
[32] Janice K. Jensen, "Pesticides Donations and the Disposal Crisis in Africa," *Pesticides News*, no. 14 (December 1991): 5–6.
[33] Pesticides Trust, "Hazardous Pesticide Dumps in Africa," *Pesticides News*, no. 14 (December, 1991): 3–4. Banned and obsolete pesticides have been identified in about twenty African countries. The Food and Agriculture Organization estimates that the total figures for Africa run into the tens of thousands.
[34] Paul Hagan and Robert Housman, "The Basel Convention," in *The Use of Trade Measures in Select Multilateral Environmental Agreements*, ed. Housman et al. (Geneva: UNEP, 1995), 132.

countries as the waste contamination effects begin to show themselves over the long term. These effects are seen in the diminished health of people from high rates of cancer and reproductive problems, soil contamination resulting in lower agricultural productivity of the land, contamination of the food chain and of groundwater, as well as harm to wildlife and biodiversity.[35] Some more specific health effects of hazardous waste exposure include leukemia, kidney cancer, and respiratory disorders, as indicated in Box 2.1.[36]

The Political Response: Negotiation of the Basel Convention

When the extent of the international toxic waste trade with developing countries was widely publicized by environmental groups and the media in the late 1980s, there was public outcry. This concern prompted action to establish international mechanisms to control exports of hazardous waste to developing countries. In the 1980s there were no universally agreed international rules on the international trade in toxic waste. Some countries had begun to implement national laws on the trade, mainly the OECD countries that sought to protect themselves from unwanted imports. Since the late 1980s several international agreements have been reached regarding the need to regulate the waste trade internationally, as well as regionally.

In the negotiation of these agreements, nonstate actors played key roles alongside the more traditional state actors. The importance of these actors can be linked to the fact that the trade in toxic waste itself is closely tied to private economic activity in the global economy and not to state actions. Private economic actors such as industry and waste dealers were key in creating the problem. Environmental NGOs quickly formed campaigns around the issue. These campaigns had three main goals, all of which were related. First, they aimed to raise awareness about the issue among the general public. Second, they sought to stop waste traders by embarrassing them through a public awareness campaign. And third, they tried to influence state negotiations on an international regulatory framework. In this latter role environmental NGOs took key roles in formulating the text of the global agreements.

---

[35] See, for example, British Medical Association, *Hazardous Waste and Human Health* (Oxford: Oxford University Press, 1991), 93–138; Allen 1992, 206–11; Third World Network 1989, 51–53; Hilz 1992, 54–63.
[36] Hagan and Housman 1995, 132.

The international regulation of the toxic waste trade began in the early 1980s, when several international organizations began to establish rules for hazardous waste management and its trade across borders. The United Nations Environment Programme began in 1982 to draw up the Cairo Guidelines on Environmentally Sound Management of Hazardous Wastes, which were completed in 1985 and approved by the UNEP Governing Council as a nonbinding set of guidelines in 1987. The European Community (EC) and OECD each established regulations on hazardous waste movement across their own borders in the mid-1980s. The OECD in 1984 adopted a Decision and Recommendation on the Transfrontier Movements of Hazardous Wastes which was a binding agreement on OECD states regarding the trade in hazardous wastes among OECD members. In the EC, the 1984 Directive on Transfrontier Shipment of Hazardous Waste was adopted, which included binding rules on the trade in hazardous wastes among EC states. In 1986 the EC and OECD both amended these rules to apply to the export of wastes to third countries. These various sets of regulations all had in common a reliance on the principle of prior notification as a main provision. This principle stipulates that senders of waste must first inform importing countries in writing of their intended exports, and the importing country must give its consent before shipments are sent. Although these various sets of guidelines were in existence in the 1980s, they did not constitute a globally agreed-upon set of rules regarding the international waste trade. In 1985 the OECD began to draft a treaty for the control of transboundary movements of hazardous wastes between OECD member states.[37]

When the Cairo Guidelines were approved in 1987 by the governing council of UNEP, Senegal, Switzerland, and Hungary proposed that the executive director of UNEP be requested to start drafting a global convention to be based on similar principles. This proposal was ratified by the UN General Assembly, and UNEP played a key role as the organizing agency for the negotiations.[38] Greenpeace officially launched its campaign to end the waste trade in 1987, at the same time these negotiations began. Five working group meetings were held between early 1988 and early 1989 to prepare the text of the global waste trade treaty, known as the Basel Convention. Ninety-six countries participated in one or more of these

---

[37] Details on these various regulations and directives can be found in Katharina Kummer, *International Management of Hazardous Wastes: The Basel Convention and Related Legal Rules* (Oxford: Clarendon Press, 1995), 38–39, 126–71.
[38] Moctar Kebe, "Waste Disposal in Africa," *Marine Policy* 14, no. 3 (1990): 252. Roelants du Vivier 1990, 265–67; Hilz and Radka 1991, 56.

working groups, of which sixty-six were from developing countries. Also participating as observers were four UN bodies, eight intergovernmental organizations, and twenty-four NGOs representing both environmental groups and industrial interests.[39] When work on the Basel Convention was nearly complete, the OECD suspended work on its waste trade treaty because there was significant overlap between the two agreements.[40]

The working group meetings leading up to the Basel Convention were very politically charged because there were basic differences of opinion on how to reconcile the exchange of hazardous wastes with the principle of free trade.[41] Two opposing viewpoints quickly emerged. On one hand, there were those who wanted the waste trade across borders to continue to be legal. Waste dealers and waste-producing firms that were reaping large profits on such deals obviously wanted to have few restrictions on their activities. Some of these firms were represented at the Basel negotiations by business advocacy groups such as the International Chamber of Commerce and the International Precious Metals Institute. Similarly, the states in which most wastes are generated, mainly the rich industrialized countries, advocated regulation of the waste trade rather than a ban. They claimed that they wanted to keep their waste management options open, even if this included the export of hazardous wastes. Such a position was clearly intended to protect the interests of powerful waste–producing firms located within their borders. UNEP agreed that the trade should be regulated rather than banned outright. It argued that not all states are able to dispose of their wastes safely and that they need to export them to countries that could do a better job at disposal. But UNEP also expressed its view that waste should not be exported to developing countries that did not want to import it. It saw the main purpose of the convention as protecting the rights of developing countries to refuse waste imports.[42]

The less industrialized countries were strongly in favor of an outright global ban on waste exports from rich to poor countries. These states saw regulation as merely legalizing of a conspicuously unjust practice. Some developing country governments were initially involved in waste trade deals, but when the widespread nature of the problem became evident,

---

[39] Mostafa Tolba and Iwona Rummel-Bulska, *Global Environmental Diplomacy* (Cambridge, Mass.: MIT Press, 1998), 112. For background on the negotiations, see Katharina Kummer, "The International Regulation of Transboundary Traffic in Hazardous Wastes: The 1989 Basel Convention," *International and Comparative Law Quarterly* 41, no. 3 (1992): 534.
[40] Kummer 1995, 161.
[41] Wynne 1989, 123.
[42] Tolba 1990, 207–8.

there was outcry among developing country governments that the export of toxic wastes was yet another mechanism to exploit them. These governments saw the negotiations as an ideal forum to demonstrate solidarity.[43] The key interest of these states was to preserve not just the environment but also justice and economic development prospects over the long term. President Gnassingbe Eyadema of Togo, referring to the Basel Convention, explained, "Our efforts for the economic development of our states and for the progress of our people will be in vain if we do not ... preserve the lives of our people and the environment."[44] Also advocating a ban on the waste trade between rich and poor countries were environmental NGOs, the most active being Greenpeace International. This group argued that as long as rich countries could legally continue to pay poor countries to take their toxic by-products, there would be no incentive to adopt clean production methods. The main tactics that this group pursued were global lobbying at Basel meetings, raising public awareness through demonstrations, and publishing extensive research reports on specific cases, naming firms and countries involved, to reveal the extent of the trade.

Once the African waste import schemes came to light, several African governments met in May 1988 at a regional workshop on toxic waste held in Monrovia, Liberia. Participating were West African governments, United Nations experts, and representatives of various environmental NGOs. This workshop recommended banning the movement of toxic waste to Africa and the elaboration of a regional convention on the waste trade. Also discussed was the establishment of national committees on environmental protection, the harmonization of environmental legislation, and the establishment of a Third World Environment Bureau to house a data bank on toxic waste issues.[45] Shortly after this workshop, the Organization of African Unity (OAU) held its forty-eighth ordinary session in Addis Ababa. At this meeting African governments adopted Resolution 1153, which strongly condemned those involved in the import of waste, and declared the practice a crime against Africa and the African people.[46] The forcefully worded resolution, adopted just as many African waste deals were being exposed, also called upon African countries that had accepted hazardous wastes, or were in the process of doing so, to cease such

---

[43] Kummer 1992, 535–36.
[44] Cited in *Waste Trade Update* 1, no. 2 (1988): 1.
[45] Kebe 1990, 251–52.
[46] Organization of African Unity, *Resolution on Dumping of Nuclear and Industrial Waste in Africa*, CM/Res. 1153 (Addis Ababa: OAU, 1988).

contracts.[47] The following month the issue was discussed at the eleventh summit of the Economic Community of West African States (ECOWAS) in Lomé, Togo. At this summit, ECOWAS leaders adopted a resolution in which they denounced the waste trade and pledged to adopt national legislation that outlawed the acceptance of foreign wastes.[48]

This heightened interest of African countries in the Basel negotiations aroused concerns that these countries might take action to block agreement on the final text of the Basel Convention. This fear prompted UNEP to agree to hold an African Ministerial Conference in Dakar in January 1989. The purpose of this conference was to encourage the African countries to agree on a common position with the developed countries before the March 1989 Basel meeting at which the convention was scheduled to be adopted. But little was agreed upon at this conference. Some representatives from industrialized countries tried to force African governments to accept prior notification as the foundation of the Basel Convention. This pressure put off many African countries, who by then were demanding a ban on exports to Africa in return for their support for the convention. The African Ministerial Conference was in the end only able to pass a broadly worded appeal for the participation of African states in the Basel negotiations. Moreover, it was clear that the African states would go ahead with their own regional convention.[49]

Between mid-1987 and mid-1988 similar moves in favor of regional agreements banning waste imports were made by the states of the Zone of Peace and Cooperation in the South Atlantic, the Non-Aligned Movement (NAM), and the Caribbean Community (CARICOM).[50] These regional coalitions soon joined forces to call for a ban on the waste trade with developing countries. These groups felt strongly that waste should not be treated as a regular commodity subject to the principle of free trade, or even regulated trade.

Environmental NGOs were important in influencing these positions taken by developing countries. Because they were both on the same side of the issue, a strong alliance was formed between environmental NGOs and Third World, particularly African, negotiators to press for inclusion of a

---

[47] Countries in the process of completing waste deals at the time included Guinea-Bissau, Benin, Congo, and Somalia.

[48] ECOWAS, *Resolution of the Authority of Heads of State and Government Relating to the Dumping of Nuclear and Industrial Waste*, A/Res.1/6/88 (1988). Although this group was formed, its activities were limited because of a lack of funding.

[49] UNEP, *Final Joint Declaration of the Dakar Ministerial Conference on Hazardous Wastes* (Geneva: UNEP, 1989), 26–27. See also Kummer 1992, 536; Kebe 1990, 252; Tolba and Rummel-Bulska 1998, 109–10.

[50] See Vallette and Spalding 1990, 18, 64, 121.

ban on the waste trade in the convention. Throughout the Basel Convention negotiating process, the NGOs that formed part of this Third World–NGO alliance were major players. Greenpeace was the main coordinator of the environmental NGOs on the issue, and its representatives began to attend the Basel working group meetings in mid-1988. At the same time it began to publish a quarterly newsletter to inform the public and interested governments of waste trade deals around the world.[51] The extensive on-the-ground research by Greenpeace on this issue gave it an expertise unmatched by most states, other NGOs, and possibly even UNEP. Other international environment and development NGOs concerned about the Basel negotiations also linked up with Greenpeace to form a temporary lobbying group, International Toxic Waste Action Network (ITWAN), to strengthen the campaign for a global ban.[52]

The Group of 77 developing countries (G-77) looked to these NGOs, Greenpeace in particular, for vital information and help in writing up proposals to be included in the convention. They also sought advice on negotiation strategy. Although the environmental NGOs were observers rather than full voting participants, they managed to wield significant influence at the negotiations through their close relationship with representatives of the developing countries. Some have argued that the concerns of the G-77 countries, such as the call to minimize waste generation and for more stringent disposal standards, were clearly shaped by the agenda of the environmental NGOs.[53] For this reason, some industrialized states began to regret that they had allowed environmental NGOs into the negotiating process. They began to close certain meetings in order to keep them out of discussions on highly sensitive issues. But developed country representatives soon realized that they could not keep the environmental NGOs from finding out what happened in those meetings because the developing countries would immediately brief them. The industrialized states and industry groups, though not organized into as tight a coalition as the G-77 and environmental NGOs, still wielded substantial power over the outcome of the negotiations. The industrialized states in effect threatened not to become parties to the convention if it went

---

[51] This newsletter was called *Waste Trade Update*. It was renamed *Toxic Trade Update* in 1992 and renamed *International Toxics Investigator* in 1995. Its publication ceased in 1997.

[52] These NGOs included African NGOs' Environment Network, Environment Liaison Centre, Greenpeace, International Organization of Consumers' Union, and the Natural Resources Defense Council. See *Waste Trade Update* 2, no. 1 (1989): 3. This network was dissolved after the adoption of the Basel Convention, and Greenpeace took over as the main NGO active in attempts to strengthen the Basel Convention.

[53] Willy Kempel, "Transboundary Movements of Hazardous Wastes," in *International Environmental Negotiation*, ed. Gunnar Sjostedt (London: Sage, 1993), 52.

so far as to ban the trade. This was a serious threat. If the major waste producing and exporting states were unwilling to ratify the treaty, there was little point to the convention.

The Basel Outcome

UNEP had set an ambitious schedule of just over a year to complete the Basel negotiations. Representatives of 116 governments and several NGOs were present at the meeting in March 1989 when the convention was to be adopted. Throughout the meeting there was a high degree of uncertainty as to whether any agreement would be reached. The end result of UNEP's initial efforts was the Basel Convention on the Transboundary Movement of Hazardous Wastes, narrowly agreed upon in the last remaining hours of the meeting.[54] The principal objectives of the Basel Convention are stated as being the reduction in the generation and transport of hazardous wastes and the promotion of the environmentally sound management of hazardous wastes, including disposal as near as possible to the source.[55] To facilitate the achievement of these objectives, the provisions of the original 1989 convention provided a regulatory framework for the movement of wastes. This regulatory framework did not constitute a comprehensive ban on their movement between rich and poor countries. The convention was later amended, in 1995, as will be discussed in Chapter 3. The analysis here focuses on the original 1989 convention.

The convention first outlines what wastes are and are not governed by the convention. Wastes exhibiting certain "hazardous" characteristics are listed in specific annexes to the convention. Annex I lists hazardous waste stream categories, and Annex III lists hazard characteristics. Trade in Annex I wastes is governed by the convention, unless they do not exhibit any of the qualities listed in Annex III. Wastes not defined as hazardous by the method above, but which are considered hazardous by either the state of export, import, or transit, are also covered by the convention.[56] Radioactive wastes are outside of the scope of the convention.[57]

[54] Kummer 1992, 537.
[55] See Kummer 1995, 47–48; Tolba and Rummel-Bulska 1998, 114–15.
[56] *The Basel Convention on the Transboundary Movement of Hazardous Wastes and Their Disposal*, Article 1.
[57] Radioactive wastes were excluded at the insistence of the International Atomic Energy Agency (IAEA), the UN organization responsible for dealing with radioactive products and technology transfer. At this time the IAEA did not have a set of regulations for the transfer of radioactive wastes, but it did begin to draft a set of regulations similar to the Basel Convention which were completed in 1991. See Patricia Birnie and Allen Boyle, *International Law and the Environment* (Oxford: Clarendon Press, 1992), 335.

A key foundation of the convention is that it affirms that countries have the sovereign right to ban imports of hazardous waste if they so choose. Parties to the convention are prohibited from exporting wastes to states that have banned its import. The convention includes restrictions on waste trade in other instances as well. One of these is the outright prohibition of exports of hazardous wastes to Antarctica.[58] It also stipulates that parties are to refrain from trade in hazardous wastes with nonparties unless there is a bilateral or regional agreement under which wastes are to be disposed of in no less environmentally sound a manner than that outlined in the convention (Article 11).[59] The purpose of this latter trade restriction is to encourage states to become parties to the convention, though the allowance of bilateral or multilateral agreements does ease this pressure somewhat.[60]

For exports of wastes to states that have not prohibited their import, the principle of prior notification is to be adhered to (Article 6). This system was intended to allow countries to accept or turn away such imports on a case by case basis at their own discretion. Parties planning to export hazardous wastes must notify the importing country in writing in advance of a hazardous waste shipment. If the latter agrees in writing to accept the waste and is assured that the waste will be disposed of in an environmentally sound manner, the transaction can take place.[61] Though these rules do not constitute a ban on the trade in wastes between rich and poor countries, the convention does call for parties to undertake a reevaluation of a ban. This measure calls on parties to undertake, three years after entry into force and at least every six years beyond that time, "an evaluation of its effectiveness and, if deemed necessary, to consider the adoption of a complete or partial ban of transboundary movements of hazardous wastes and other wastes in light of the latest scientific, environmental, technical and economic information."[62] This measure was included in response to a proposal made by Greenpeace.[63]

Beyond these specific trade provisions regarding the movement of hazardous waste, the convention outlines other steps that should be taken to meet its stated objectives. Perhaps most important, it requests parties to take steps to reduce the generation of hazardous wastes and to ensure the

---

[58] *The Basel Convention*, Article 4.
[59] Ibid., Articles 4, 11.
[60] For a full analysis of the trade measures in the Basel Convention, see Krueger 1999.
[61] *The Basel Convention*, Articles 4, 6.
[62] Ibid., Article 15.7.
[63] Ibid., Article 15; Kummer 1992, 539.

availability of environmentally sound disposal facilities. The transboundary movement of hazardous wastes, though allowed, is to be kept to a minimum. The convention also requests parties to export wastes only if they themselves lack the means to dispose of those wastes in an environmentally sound manner or if the wastes are considered a "raw material" by the importing country.[64] Movements carried out in contravention of provisions of the treaty are considered to be illegal traffic in waste, and the waste must be reimported by state of export.[65] The convention established a secretariat to arrange periodic conferences of the parties (COPs) at which all contracting parties would meet. The secretariat is to act as a liaison center for information on waste management, technical assistance, and the identification of illegal trade in wastes.

The final text of the Basel Convention incorporated some of the demands of the environmental NGOs and G-77 states. These include the need to reconsider the issue of a ban in light of new evidence and the incorporation of provisions calling for environmentally sound disposal of exported wastes. But the convention was an overall disappointment to the environmental NGOs and developing countries because it was based on regulating the trade in hazardous wastes, not banning their export. It was also a disappointment to some of the key industrialized countries and industry groups, who thought that the convention's provisions went too far in regulating the trade in wastes. The battle that emerged during the negotiation of the convention threatened to stall its acceptance on a global scale.

Although some proposals made by African countries during the negotiations were incorporated into the final document,[66] the African country delegates were not happy with the outcome. Although thirty-eight of forty-three African countries at the final conference adopted the convention in principle, all suspended signature of the document.[67] As Mostafa Tolba has noted about this meeting, "The problem of the African delegates' intransigence hung heavy in the air."[68] The African government representatives said that they could not sign or ratify the Basel Convention until they consulted with one another at the upcoming OAU meeting to be held in June 1989, at which they would discuss drafting their own con-

---

[64] *The Basel Convention*, Article 4.9.
[65] Ibid., Article 9.
[66] UNEP, *Proposals and Position of the African States during the Negotiations on the Basel Convention on the Control of Transboundary Movements of Hazardous Wastes and Their Disposal and the Status of Their Incorporation into the Basel Convention* (Geneva: UNEP, 1989).
[67] Kebe 1990, 252.
[68] Tolba and Rummel-Bulska 1998, 112–13.

vention. Tolba directly attributes the African delegates' refusal to sign the treaty at Basel to Greenpeace. In pushing for a global ban on the waste trade and highlighting the ways in which the Basel Convention worked against this goal, Tolba remarked that "the organization convinced some African representatives, who later blocked the rest of the African states from signing the convention when it was finally adopted."[69] Other developing countries also withheld signature of the document for similar reasons. Environmental NGOs immediately denounced the agreement as the legalization of "toxic terrorism."

Nonetheless, the final text of the Basel Convention was signed by thirty-five countries in March 1989 and was ratified by the necessary twenty countries to come into force by May 1992. This was accomplished largely without the participation of developing countries[70] or of the EC (except France), the United States, and Japan. This latter group of countries together produce over 90 percent of hazardous waste.[71] The Basel Convention thus got off to a fragile start. The key players involved in the waste trade negotiations at the time, the industrialized countries and waste traders on the one hand, and the developing countries and environmental NGOs on the other, were, for opposite reasons, disappointed with the treaty.

National Legislation and Regional Waste Trade Agreements

Following the negotiation of the Basel Convention in 1989, several countries and regional organizations in the developing world began to take other precautionary measures to control the waste trade. Encouraged by Greenpeace and local environmental NGOs, many countries began to implement national legislation banning the import of hazardous waste. The number of countries banning the trade rose from 3 in 1986 to 103 by early 1994.[72] These countries include most developing countries, as well as several industrialized countries, including Italy and Norway. The latter two

---

[69] Ibid, 103.
[70] Nigeria, however, was one of the first twenty countries to ratify the Basel Convention. It went against OAU instructions not to ratify the treaty until the Bamako Convention had been ratified. It is unclear why Nigeria has taken this position, although it has been rumored that its decision was linked to pressure from certain aid donors.
[71] The United States produces 85 percent of the world's hazardous wastes, while the EC produces 5–7 percent of the world total. *Environmental Policy and Law* 23, no. 1 (1993): 14.
[72] Puckett 1994, 55.

had each enacted a ban in the late 1980s following incidents of dumping that originated from those countries.

In 1989, directly following the Basel negotiations, the African, Caribbean, and Pacific (ACP) states insisted that the EC (now the European Union, EU) impose a ban on exports of hazardous wastes as well as radioactive wastes to the ACP states within the framework of the Lomé IV Convention. The Lomé Convention is an aid and trade agreement between the European Union and the ACP states which is renegotiated on a regular basis. In response, the EU proposed to allow the export of hazardous wastes to ACP countries if accompanied by technologies to improve waste disposal safety and management.[73] The ACP states refused to accept this proposal and demanded an outright ban on their export to ACP states. A compromise was reached whereby the ACP states agreed not to accept toxic waste exports from any country in return for a ban on exports from the EU states. This agreement was outlined in Article 39 of the Lomé IV Convention.[74] The Lomé IV agreement came into force in 1991 and was valid until the year 2000. Because of new global trade dispute rules under the WTO which came into effect in 1995, it is unlikely that a new Lomé agreement will be negotiated. This will not affect the EU's ban on the export of wastes to ACP states because the EU has subsequently banned the export of hazardous wastes to all non-OECD countries.

Seeking to protect themselves from waste exports from non-EU states, the African countries agreed at the June 1989 OAU meeting to draft an African waste trade convention. Three working group meetings were held to prepare and draft the convention. Present at these meetings were African ministers of the environment and foreign affairs, as well as legal and technical experts. The latter included a technical expert from UNEP and a legal expert from Greenpeace. UNEP officials initially felt that a separate African convention might undermine the Basel Convention. At these negotiations the UNEP representative tried to convince the Africans not to abandon the Basel Convention and suggested that they issue a declaration of their concerns instead.[75] But the Africans were determined to have their own convention and continued to work toward that goal. The Bamako Convention on the Ban of the Import into Africa and the Control of Transboundary Movement and Management of Hazardous Wastes

---

[73] Hilz 1992, 153.
[74] EEC-ACP, *Fourth ACP-EEC Convention*, signed in Lomé on December 15, 1989 (Article 39), 18.
[75] Interview with Kevin Stairs, Greenpeace International, November 1993; UNEP, *Hazardous Waste: Why Africa Must Act Now* (Geneva: UNEP, 1989).

within Africa was completed and signed by twelve African countries at the OAU Pan African Coordinating Conference on Environment and Sustainable Development held in Bamako, Mali, in early 1991.[76]

The text of the Bamako Convention closely mirrors that of the Basel Convention in some respects, but for the African countries it is an important improvement on the latter. It imposes an outright ban on the import of hazardous wastes, including radioactive wastes, into African countries.[77] The Bamako Convention also bans all forms of ocean dumping of wastes; outlaws the import of hazardous substances that have been banned in the country of manufacture; and includes provisions on clean production methods within Africa. It further requires hazardous waste audits and imposes strict, unlimited, joint, and several liability on waste generators.[78] The inclusion of more stringent provisions than the Basel Convention has been attributed to the influence of Greenpeace.[79] The African states recognized that the Bamako Convention would lack the funding to monitor effectively the movement of toxic wastes and called on NGOs to assist in ensuring compliance. They also acknowledged the important role NGOs played in the drafting of the convention. The executive director of the Kenya Energy and Environment Organization, Achoka Awori, said of the convention: "It sends a message even though it doesn't do much. It gives environmental agencies something to point to in a legal sense.... It is now up to the NGO community and the international agencies to patch up the loopholes."[80] The Bamako Convention gained the necessary ten ratifications and came into force in March 1996.

The Agreement on the Transboundary Movement of Hazardous Wastes in the Central American Region was signed by six Central American presidents in Panama in late 1992. This convention was adopted following a year of strong campaigning by environmentalists in the region. A number of NGOs had formed the Central American Committee against Toxic Waste Trafficking and Other Polluting Products to monitor the waste trade in the region and to lobby Central American governments. This group was able to convince these governments to adopt a regional

---

[76] OAU, "Africans Ban Hazardous and Nuclear Waste Dumping," press release, January 29, 1991.
[77] *Bamako Convention on the Ban of the Import into Africa and the Control of Transboundary Movement and Management of Hazardous Wastes within Africa* (Addis-Ababa: OAU, 1991).
[78] Ibid.; see also "Africa Adopts Sweeping Measures to Protect Continent from Toxic Terrorism," *Waste Trade Update* 4, no. 1 (1991): 1.
[79] Interview with Kevin Stairs, Greenpeace International, November 1993.
[80] "United Nations Officials See Basel Treaty as 'Limping' into Effect with Limited Support," *International Environment Reporter* 15 (May 6, 1992): 275–276.

waste trade ban in addition to the existing national bans.[81] The resulting convention is similar to the Bamako Convention in that it calls for a ban not only on waste imports to the region but also on transportation, ocean dumping, and ocean incineration of hazardous wastes in Central America. The agreement is valid for ten years, after which time it must be renewed. Governments seeking to opt out can do so on six months' notice.[82] This agreement is currently in force in the six Central American countries that signed it.

Nineteen countries in the Mediterranean region agreed in late 1993 to negotiate a protocol banning the waste trade within the framework of the existing Barcelona Convention for the Protection of the Mediterranean Sea Against Pollution (1975). This agreement came following a decision in late 1991 of the parties to the Barcelona Convention to establish a working group of experts to draft this protocol.[83] The result was the Protocol on the Prevention of Pollution of the Mediterranean Sea by Transboundary Movements of Hazardous Wastes and Their Disposal, known as the Izmir Protocol, which was adopted October 1, 1996, in Izmir, Turkey. The protocol calls for protecting of the Mediterranean Sea from hazardous waste by banning the trade and transit of hazardous wastes and their disposal (including wastes destined for recycling) between industrialized and developing countries in the region.[84] According to Katharina Kummer, "This draft had been prepared by Greenpeace International and reviewed by the competent UNEP institutions."[85] This protocol is not yet in force.[86]

Also in late 1993 the South Pacific Forum began negotiations on a regional convention known as the Waigani Convention. The official title of the agreement is The Convention to Ban the Importation into Forum Island Countries of Hazardous and Radioactive Wastes and to Control the Transboundary Movement and Management of Hazardous Wastes within the South Pacific Region. It was adopted September 16, 1995, in Waigani, Papua New Guinea. This convention bans the import of hazardous and radioactive wastes from outside the convention area to developing countries in the convention area, while prohibiting New Zealand and Australia from

---

[81] "Latin America Blocks Hazardous Import Schemes," *Waste Trade Update* 5, no. 1 (1992): 4.
[82] "Toxic Trade Ban Agreed by Central American Presidents," *Toxic Trade Update* 6, no. 1 (1993): 5.
[83] *Siren*, no. 45 (March 1992): 22–24.
[84] "Three Regions Move to Ban the Waste Trade," *Toxic Trade Update* 6, no. 4 (1993): 4.
[85] Kummer 1995, 120.
[86] As of March 31, 1999, only one country, Tunisia, had ratified this protocol.

exporting hazardous or radioactive wastes to the developing countries in the convention area.[87] This convention is not yet in force.

Other regional efforts were made in the early 1990s, for example within the Association of Southeast Asian Nations (ASEAN) and in the Economic Community of Latin American Countries (ECLAC), but these agreements have not yet been completed.[88]

Conclusion

The initial rise of the waste trade is best explained as a combination of both local economic incentives and the nature of the global economy. Increased regulations raised costs for hazardous waste disposal in rich countries, making waste exports to developing countries with less stringent regulations and lower costs appealing. The lure of foreign exchange made waste imports attractive in poor countries. While these factors are crucial in explaining the emergence of the trade, without a fluid global trading system, reinforced by lower costs for transportation and communications, it would not have been lucrative or easy. The global factor in the equation, one which is often overlooked or underplayed, is extremely important. This set of circumstances led to a rapid growth in attempts to export toxic waste to developing countries in the late 1980s. Though not all of the attempts resulted in actual waste dumping once they were exposed by NGOs and the media, the sheer number of proposals indicates that the problem could have easily exploded into a major crisis had it not been put in the international spotlight.

The political response to the rise of the waste trade was the negotiation of the Basel Convention as well as other regional waste trade agreements. In the formulation of these rules, states played an important role as the official participants in the negotiations and as official parties to the treaties. The relative power position of states in the negotiations was indeed important in determining the final outcome of the Basel Convention and other agreements. But they were far from being the only significant players in international waste trade politics. Alongside state actors, nonstate actors played key roles in the formulation of the Basel waste trade convention.

---

[87] "South Pacific Forum Countries Sign Regional Hazardous Waste Convention," *International Environment Reporter* 18, no. 19 (September 20, 1995): 709–10. See also "South Pacific Forum to Negotiate a Regional Waste Trade Ban," *Toxic Trade Update* 6, no. 3 (1993): 4.
[88] "Southeast Asian Activists Call for a Regional Waste Trade Ban," *Toxic Trade Update* 6, no. 3 (1993): 5.

Both environmental NGOs and business lobby groups were active in the process by lobbying relevant government representatives. Greenpeace International in particular participated actively as an independent negotiator. Greenpeace representatives were also indirectly influential as advisers to developing country governments at the Basel negotiations, providing them with technical information on waste dumping and helping them to formulate strategy. The alliance that formed between Third World governments and Greenpeace was mutually beneficial, as both had a significant amount to gain from cooperation. While the developing country governments gained access to vital information and strategy advice for negotiations, Greenpeace gained influence in shaping the direction of those negotiations by operating through developing country government representatives when meetings were closed to NGOs. Greenpeace also played a strong role, perhaps stronger than at the Basel meetings, in the drafting of many of the regional waste trade agreements. Industry interests were also present during the negotiation of the Basel Convention, strongly lobbying in opposition to strict regulation on the movement of wastes. These players were aligned in their views with the industrialized countries, though they did not form as tight an alliance as did the environmental groups and developing countries. As one observer notes, environmental NGOs were more directly vocal and influential over the final wording of the Basel Convention than were industry lobby groups.[89]

The factors that initially led to the transboundary movement of wastes are not unrelated to the importance of nonstate actors in the political response to it. The global economic factors that contributed to the rise of the waste trade also allowed for the rise in nonstate actors as key players in the global waste trade negotiations. Moreover, the fact that the waste trade was conducted by private, rather than state actors, helps to explain why environmental NGOs were among the first to identify the practice and to publicize it. Their work subsequently focused on states in an attempt to get them to hammer out an acceptable international agreement to govern the trade. But their early involvement in identifying and publicizing the issue with the broader public secured them a place at the negotiating table.

[89] Kempel 1993, 51.

3
The Role of Environmental NGOs in the Evolution of the Basel Ban

The Basel Convention and other regional and national regulatory measures were specifically aimed at reducing the export of hazardous wastes to developing countries. In this goal they were partly successful. The number of proposals for the export of toxic waste to developing countries for final disposal indeed declined dramatically after the adoption of these agreements. But by the mid-1990s, it was clear that they had weaknesses that enabled waste traders to circumvent the various treaties' rules. Waste exporters responded to the agreements by shifting from the export of hazardous wastes destined for final disposal to that destined for recycling operations, which made the waste trade more difficult to track. Though recycling implies environmental stewardship of hazardous wastes, in most cases, especially in developing countries, it has been as harmful to the environment as dumping. Another response of waste traders was to label wastes as other products, again making them difficult to monitor. The rapid growth in the hazardous waste exports to developing countries for purposes of recycling or further use represents another dimension of hazard transfer. This practice was in large part a response to the implementation of international regulations on toxic waste exports, and it was facilitated by the fluid nature of the global economy.

The political response was swift. Environmental NGOs were quick to point out the weaknesses in the regulatory framework that enabled waste traders to circumvent the growing number of international rules governing the waste trade. Their strategy was to continue to pressure parties to require a total ban on the waste trade between rich and poor countries, including the shipment of wastes for both disposal and recycling, in the Basel Convention. Such a ban would place the burden of stopping unwanted waste shipments not just on importing states but also on exporting states. The strategy of the NGOs in their attempt to achieve this goal was to embarrass those states that opposed such a ban. The battle for the ban was a long and drawn-out process, in which the issue of recycling was the most contentious aspect.

This chapter addresses the weak points of the initial international efforts to stop the waste trade, focusing in particular on the slow implementation and legal weaknesses of the Basel Convention. I discuss the major weakness in all of the waste trade agreements, that of exporting wastes under the name of recycling and further use and its impact on developing countries. Environmental NGOs carried out extensive research on this phenomenon as part of their campaign to push for a complete ban on both disposal and recycling in the context of the Basel Convention. Agreement on such a ban was finally secured in the mid-1990s.

Legal Weaknesses of the Basel Convention

As with other international environmental agreements, the Basel Convention is binding only on those states that are parties to it, meaning that they have ratified or acceded to the agreement. To do this, states must implement national legislation that conforms to the requirements of the treaty. The treaty comes into force only when a certain number of states have ratified it. In the case of the Basel Convention, twenty ratifications were necessary before the convention came into force. Despite widespread interest at the Basel Convention's adoption in 1989, and the fact that more than one hundred countries signed the document at that time, it did not come into force until May 1992.

Of the six largest waste-producing countries (United States, Canada, United Kingdom, Australia, Japan, and Germany), only Canada and Australia ratified the convention before the end of 1992. Japan ratified in late 1993, and the EU states, which include the United Kingdom and Germany, ratified in mid-1994.[1] The United States expressed early on that it was also interested in ratifying as soon as possible so it would be able to take part in the process of defining the convention's provisions at the conferences of the parties. But progress on the adoption of implementing legislation has been extremely slow. Under the George Bush Sr. administration, several bills that would enable the government to ratify the Basel Convention were debated in the U.S. Congress, but were not passed into law. Hopes were raised that swift action would be taken when President Bill Clinton came to office because Vice President Al Gore was strongly in favor of a global waste trade ban.[2] But the Clinton administration did not follow through with its intentions to ratify the treaty by the end of 1993.[3] Further developments in the convention regarding a partial ban on the waste trade between rich and poor countries brought U.S. activity toward the ratification of Basel to a standstill for nearly five years.

Many of the developing countries also delayed ratifying the convention, which left them vulnerable to waste exports from states that were not parties to the Basel Convention. African countries in particular delayed ratification because they had chosen to focus instead on the Bamako Convention. Similarly, other regions delayed ratification in the hopes of developing regional agreements that would embody stronger language than the Basel Convention. Developing countries were in fact discouraged by Greenpeace from ratifying the convention at that time because of its failure to incorporate a ban. Because of this delay from the developing countries, the then executive director of UNEP, Mostafa Tolba, appealed to these states, particularly the Africans, to ratify the Basel Convention as well as regional agreements.[4]

The main substantive weakness of the 1989 Basel Convention in the eyes of its opponents was that it did not include an outright ban on the

---

[1] See "United Nations Officials See Basel Treaty as 'Limping' into Effect with Limited Support," *International Environment Reporter* 15 (May 6, 1992), 278; "Dirty Half-Dozen Stand Alone," *Toxic Trade Update* 6, no. 3 (1993): 3.
[2] Indeed, Al Gore was a member of GLOBE (Global Legislators for a Balanced Environment), which advocated a ban on the trade with developing countries.
[3] "Dirty Half-Dozen Stand Alone," *Toxic Trade Update* 6, no. 3 (1993): 3–4.
[4] "UNEP Official Urges African Nations to Approve Basel Accord on Waste Shipments," *International Environment Reporter* 15 (October 7, 1992): 654.

waste trade between rich and poor countries. Many environmental groups and developing countries feared that the regulation of the waste trade by the Basel Convention would not halt the waste trade with those countries because the obligations of the parties are set out in very ambiguous language. Noncompliance with the treaty would be both easy to accomplish and difficult to detect.

One of the main limitations of the Basel Convention in this respect is that key terms which are used to lay out the rules and obligations of parties are only vaguely defined. Agreement on a clearer definition of many terms proved to be politically difficult at the negotiating sessions. For example, "environmentally sound management" is defined as "taking all practicable steps to ensure that hazardous wastes or other wastes are managed in a manner which will protect human health and the environment against the adverse effects which may result from such wastes."[5] This definition is open to subjective interpretation and has thus made it very difficult to enforce the obligation of parties that wastes can cross borders only if disposed of in an "environmentally sound manner."

The original convention also provides no precise definition of "hazardous waste." The annexes that determine whether a waste is hazardous were seen by many to be open to interpretation. Although the annexes to the convention list waste streams and properties of hazardous wastes, the convention also recognizes that the potential hazards posed by certain wastes have not yet been determined. It has also been argued that toxic products, such as pesticides banned in one country but being exported to another in which they are not banned, are not included in the Basel definition of waste because they are not technically destined for disposal.[6] Adding to these ambiguities, each country has a different definition of hazardous wastes.[7] As one waste expert commented regarding the Basel definition of hazardous waste, "The imprecision of many key terms and the chronic inconsistency of hazardous waste definitions leaves the boundaries between legal and illegal, satisfactory and unsatisfactory practices ill-defined."[8]

---

[5] *The Basel Convention*, Article 2(8).
[6] Debora MacKenzie, "If You Can't Treat It, Ship It," *New Scientist* 122 (April 1, 1989): 24.
[7] *The Basel Convention* (Annex III). Some attempts have been made to harmonize the definition of hazardous wastes by the OECD and UN, but these definitions are not yet in use; see also Laura Strohm, "The Environmental Politics of the International Waste Trade," *Journal of Environment and Development* 2, no. 2 (1993): 131.
[8] Brian Wynne, "The Toxic Waste Trade: International Regulatory Issues and Options," *Third World Quarterly* 11, no. 3 (1989): 140.

The prior notification procedure on which the Basel Convention is based has also been attacked for being ineffective. The exporting state must inform and receive the consent of the importing state before shipping its wastes. But this letter of consent does not have to be sent on to the Basel Convention secretariat for inspection unless specifically requested by a party who has reason to question the agreement.[9] As a result, it is difficult to ascertain whether written notices are authorized by the proper officials or whether they are worded in such a way as to enable the importing country to assess whether or not to accept the offer.[10] The Basel Convention secretariat was not given any powers to monitor the behavior of parties or to apply sanctions to ensure compliance.[11] With such a weak mechanism for ensuring compliance, there was a worry that developing countries might be pressed into accepting waste imports with no adequate check on whether wastes are disposed of safely.[12]

A further weakness of the Basel Convention was that it allowed bilateral waste trade agreements between parties and nonparties under Article 11. Such transactions are permitted provided the relevant waste trade agreements ensure that the waste is disposed of in a manner that is at least as environmentally sound as the provisions in the Basel Convention. But the definition of "environmentally sound" is extremely vague. This ambiguity opens the door for possible waste trade agreements with developing countries that bypass the rules of the Basel Convention.[13] The existence of Article 11 agreements has been a major worry of environmental groups.[14]

Other faults with the 1989 convention identified by analysts include its lack of provisions for liability and compensation for toxic waste contamination, its failure to cover the trade in radioactive wastes, and the absence

---

[9] One analyst claims that "the notification and consent procedure has been reduced to another opportunity for bribery" (Strohm 1993, 141). See also Hilary French, "A Most Deadly Trade," *World Watch*, July–August 1990, pp. 13–15; Katharina Kummer, "The International Regulation of Transboundary Traffic in Hazardous Wastes: The 1989 Basel Convention," *International and Comparative Law Quarterly* 41, no. 3 (1992): 548–49; Wynne 1989, 140.

[10] Greenpeace International, "The Failure of Prior Informed Consent" (Attachment 'C'), in "Annotations of Greenpeace International on the Agenda of the Meeting," paper prepared for the First Conference of Parties to the Basel Convention, Piriapolis, Uruguay, November 30–December 4, 1992: 2–3.

[11] Katharina Kummer, "Minimization and Control of International Traffic in Hazardous Wastes: Ensuring Compliance," paper presented at the International Studies Association annual meeting, Acapulco, March 1993.

[12] Wynne 1989, 140.

[13] Kummer 1992, 561.

[14] Greenpeace International 1992, "A Basel Convention Overview, Key Provisions and Faults" (Attachment 'D') 1–3.

of firm incentives to eliminate the generation of hazardous waste. In addition, the legal status of the trade provisions contained in the convention with respect to the GATT agreement on international trade was not clarified when the convention was signed. Many of these weaknesses occurred because the negotiators could not reach agreement. The loose ends were to be clarified at the subsequent meetings of the technical and legal working groups and the conference of the parties.

The Growth in Toxic Waste Exports for Recycling and Further Use

Despite the legal and definitional weaknesses in the Basel Convention, waste traders in theory should have still been prevented from exporting wastes to countries that banned toxic waste imports. But the Basel Convention, as well as the Lomé IV and Bamako Conventions and the numerous regional waste trade agreements that explicitly ban the trade, were frequently circumvented even after they were put into place. Exporters of hazardous wastes have in many cases relabeled their wastes as commodities that are bound for recovery operations or for further use. According to Greenpeace, by the early 1990s some 90 percent of the hazardous waste trade shifted from that destined for disposal to that headed for recycling operations or further use.[15] The OECD cites statistics that show that the percentage of all hazardous waste exported from OECD countries for recycling purposes increased from 50.2 percent in 1992 to 58.4 percent in 1993.[16] Some examples of the export of hazardous waste to developing countries for the purpose of recycling are presented in Box 3.1.

The growth in exports of hazardous waste for recycling appears to be a direct response to the development of new international rules governing the trade in wastes. The Basel Convention states that wastes can be exported if they are required as raw materials for recycling in the country of import. These wastes are still subject to the prior notification procedure. But this rule has been difficult to enforce when hazardous wastes that are traded internationally are not labeled as wastes per se. Also contributing to the shift in waste exports towards recycling were new OECD and

---

[15] Interview with Jim Vallette, Greenpeace USA, Washington, D.C., September 1993. See also Greenpeace International, *Database of Known Hazardous Waste Exports from OECD to Non-OECD Countries, 1989–March 1994* (Amsterdam: Greenpeace International, 1994).
[16] OECD, *Transboundary Movements of Hazardous Wastes 1992–93 Statistics* (Paris: OECD, 1997).

Box 3.1   Some Examples of Exports of Hazardous Waste from OECD to Non-OECD Countries Destined for Recycling Operations, 1995–1997 (in metric tons)

1. Recycling/reclamation of metals and metal compounds

Germany exported:	584 tons to Kazakhstan (1996)—copper compounds
	263 tons to Slovakia (1996)—copper compounds
The U.K. exported:	3,000 tons to Estonia (1996)—lead and lead compounds
Norway exported:	1,611 tons to Estonia (1996)—copper compounds
Belgium exported:	128 tons to Hong Kong (1996)—wastes resulting from surface treatment of metals and plastics
	156 tons to China (1996)—wastes resulting from surface treatment of metals and plastics
Netherlands exported:	914 tons to India (1995)—zinc compounds
Japan exported:	1,140 tons to Korea (1995)—cadmium and cadmium compounds
The United States exported:	39 tons to Brazil (1997)—copper compounds

2. Wastes to be used for energy generation

Germany exported:	263 tons to Slovakia (1996)—copper compounds

3. Recycling/reclamation of organic substances not used as solvents

Germany exported:	91 tons to the Czech Republic (1995)—household waste

Sources: Secretariat of the Basel Convention (SBC), *Generation and Transboundary Movements of Hazardous Wastes and Other Wastes, 1995 Statistics; 1996 Statistics* (SBC: Geneva, May 1999 and June 1999); SBC, *Reporting and Transmission of Information under the Basel Convention for the Year 1997* (SBC: Geneva, November 1999).

Note: Not all parties report data to the Basel Convention secretariat. These are just examples and do not constitute a full list of all exports of hazardous waste destined for recycling operations from OECD to non-OECD countries.

EU regulations that sought to facilitate the export of waste for recycling among themselves.

The 1992 OECD Council Decision concerning the Control of Transfrontier Movements of Wastes Destined for Recovery Operations was adopted in March 1992 with a view to differentiating rules for trade in wastes destined for recycling among OECD countries. These rules divide wastes destined for recycling into three categories, known as "red, amber, and green."[17] Red wastes are considered the most toxic such as those containing PCBs, cyanide, and asbestos. These wastes are subject to strict prior notification procedures including written consent of the importing country before they can be exported, similar to the prior notification requirement under the Basel Convention. Amber wastes are seen to be potentially hazardous but less risky than red wastes. They include substances such as phenol and used lead-acid batteries. These amber wastes are subject to a more limited prior notification procedure. Notification of export must be given, but consent can be tacit rather than in writing. Green wastes are deemed safe and are not subject to the prior notification procedure but are governed by rules of normal commercial transactions. Included in the green list are wastes such as certain recyclable metals and nondispersible forms of metal alloys, as well as textiles, paper, glass, and some plastics.[18]

The EU in 1993 adopted a similar "red, amber, green" system of classifying wastes for recycling as part of a new waste regulation.[19] But rather than applying this system only internally among EU countries, as does the OECD system, the EU has applied the red, amber, green system to some exports outside of the EU. Exports of these wastes were allowed to countries (whether OECD or not) that were parties to the Basel Convention or that had bilateral, multilateral, or regional agreements with the EU consistent with Article 11 of the Basel Convention. This in effect meant that green category wastes could be exported to third countries without any prior notification. In theory the ACP countries at the very least should

[17] *Waste and Environment Today* 5, no. 2 (February 1992): 28–29; *Waste and Environment Today* 5, no. 5 (May 1992): 37. See also Jim Puckett, Paul Johnston, and Ruth Stringer, *When Green Is Not* (Amsterdam: Greenpeace International, 1992).
[18] OECD 1998, 9; Katharina Kummer, *International Management of Hazardous Wastes: The Basel Convention and Related Legal Rules* (Oxford: Clarendon Press, 1995), 161–62.
[19] This is the 1993 regulation on the supervision and control of shipments of waste within, into, and out of the European Community, http://europa.eu.int/eur-lex/en/lif/dat/1993/en_393R0259.html. The regulation was amended in 1997 to ban any waste exports for either disposal or recovery to any non-OECD state as of January 1, 1998, in accordance with the ban amendment to the Basel Convention.

have been protected from exports of green wastes because waste exports from the EU to those countries are banned under the Lomé Convention, but some have argued that this was not strictly enforced in the early 1990s. Moreover, some environmental groups charged that some green listed wastes are considered hazardous under the Basel Convention. These include some forms of plastics as well as non-dispersible forms of lead, zinc, and cadmium wastes.[20]

The growth in the export of wastes for recycling was mainly a response to these new regulations. Whereas the initial emergence of the waste trade was in part a response to differences in domestic regulations, this time the response was triggered by new rules at the international and regional levels. The fluid nature of global trade and investment facilitated this shift in the waste trade, as it was relatively easy for waste exporters to switch from exporting waste for final disposal to reprocessing of wastes. Waste recycling firms began to proliferate in the Third World. These firms were set up both by local entrepreneurs and with the involvement of transnational firms. The recycling of hazardous waste in some cases is an environmentally favorable option. But in many cases it is extremely polluting. The recovery of products from imported hazardous wastes in developing countries is in almost all cases carried out under very unhealthy and environmentally dangerous conditions. A large proportion of these wastes cannot actually be recycled. Hazardous by-products are also often left behind after the recycling process, leaving toxic waste that must be disposed of. Environmental groups charge that many hazardous wastes have been exported to developing countries for recycling when in fact the waste was not recycled at all. Greenpeace, in collaboration with local NGOs in several developing and Eastern European countries, carried out extensive research on the problem of hazardous waste recycling throughout the 1990s. It published many reports on such incidents to back up its arguments that this practice should be stopped.

Waste-to-energy schemes were one of the first strategies for waste exporters to recycle hazardous wastes in less industrialized countries. First developed in industrialized countries, this method of incinerating hazardous wastes to generate energy quickly spread to poorer countries. Waste-to-energy proposals were made in the late 1980s and early 1990s to

---

[20] *ENDS Report*, no. 215 (December 1992): 37; Debora MacKenzie, "Europe's Green Channel for Toxic Waste," *New Scientist* 137 (March 13, 1993): 13–14; Jim Puckett, "E.C. Establishes 'Waste Colonialism' as Law," *Toxic Trade Update* 6, no. 1 (1993): 9–10.

many countries in Africa, Latin America, and Eastern Europe.[21] Under such schemes, the imported hazardous waste is burned in incinerators, often to run a power plant. In return, the country is offered money, roads, ports, and, of course, incinerators and power plants. In Angola, for example, the government was offered projects to promote health and education in return not only for accepting wastes for incineration but also for the use of semi-desert land on which to store imported toxic wastes.[22]

The incineration of these hazardous wastes (which often contain PVC and other heavy metals) often releases toxins into the air because the temperatures of the kilns are not strictly controlled. The general lack of stringent emissions controls in developing countries has acted as a pull for waste traders seeking to offload their toxic wastes through such schemes. In 1992, for example, a German firm attempted to export toxic wastes to Egypt for recycling by incineration. When Greenpeace uncovered the plans the waste was eventually returned.[23] Reports that Namibia was considering a plan to import domestic waste from New York City to fuel a power plant in 1998 once again raised concerns that Africa is seen as a dumping ground for the rest of the world.[24] As more of these waste-to-energy schemes were reported by the media, many of them were canceled by the recipient countries.

Other recycling operations in developing countries have also proven damaging to health and the environment. One of the more infamous recycling operations was the mercury reprocessing plant in Cato Ridge, South Africa. This British-owned plant, Thor Chemicals, regularly imported mercury wastes to South Africa from the United Kingdom and the United States from the mid-1980s to the early 1990s.[25] This case highlights the weakness of the waste trade regime that stems from the fact that countries not party to the various conventions do not have to follow its rules. South Africa was not a member of the OAU until mid-1994 (which excluded it

---

[21] See, for example, Mohammed Sanussi, "Sierra Leone Scandal over Toxic Waste Power Scheme," *New African*, no. 265 (October 1989): 33; see also Jato Thompson, "Toxic Waste Merchants Offer $2b," *African Business* (March 1989): 16–17.
[22] "Angola: An Offer Luanda Just Can't Refuse?" *Africa Report* (March–April 1989): 5.
[23] "Egypt Sends Waste Back to Germany," *Toxic Trade Update* 5, no. 2 (1992): 12.
[24] Christopher Munnion, "World Sees Africa as Its Rubbish Dump," *Daily Telegraph*, July 17, 1998.
[25] See, for example, Chemical Workers Industrial Union, "The Fight for Health and Safety," in *Restoring the Land: Environment and Change in Post-Apartheid South Africa*, ed. Mamphela Ramphele (London: Panos Publications, 1991), 79–86; Jim Vallette and Heather Spalding, ed., *The International Trade in Wastes: A Greenpeace Inventory* (Washington, D.C.: Greenpeace 1990), 26–38; Tom Robbins, "Scandal at Thor Chemicals," *New African*, no. 301 (October 1992): 27.

from the Bamako Convention), nor was it party to the Lomé IV or Basel Conventions in the early 1990s. South Africa has considered mercury waste imports to be imports of raw materials rather than toxic waste.

The Thor mercury recycling plant has severely polluted the environment with mercury waste by-products and mercury-laced incinerator ash which have been dumped in landfills and into a sludge pond. The toxic wastes leaked into the soil and have made their way to a stream feeding into the Umgeni River.[26] This river runs through the Kwa Zulu homeland, and its waters are used for cooking, clothes washing, and bathing by the local population. Greenpeace-sponsored researchers found levels of mercury in this river in the early 1990s that were over one thousand times the safety limit set by the World Health Organization. High levels of mercury gases, dioxins, and other chemicals were released into the air during 1993 by the plant's mercury waste incinerator. As a result of unsafe practices and pollution emissions, nearly one-third of the workers at the plant suffered mercury poisoning. Several Thor workers were hospitalized with symptoms of mercury poisoning in 1992, and at least two workers subsequently died. According to environmental groups, others were permanently disabled.[27]

Several South African environmental NGOs and Greenpeace linked up their campaigns on the Thor case in an attempt to shut the plant down. Earthlife Africa and the Environmental Justice Networking Forum, both local NGOs in South Africa, have actively sought to change South African policy, which has allowed waste imports for both disposal and recycling. These groups regularly exchanged information with Greenpeace. Following pressure by South African NGOs and Greenpeace regarding the Thor Chemical Company, the government in 1993 banned waste imports for disposal (though not for recycling).[28] Following a government inquiry into the incident, Thor Chemicals decided to close those sections that used mercury compounds. It also announced that it was to cease importing spent mercury for recycling but stressed that this was not an admission that it had caused environmental harm. In early 1994 it was disclosed that Thor in fact had not been recycling mercury for several years but had

---

[26] According to Greenpeace, the levels of mercury in this stream were so high that its sediments could be classified as toxic waste. Fred Kockott, *Wasted Lives: Mercury Waste Recycling at Thor Chemicals* (Amsterdam: Greenpeace International and Earthlife Africa, 1994).
[27] For full details on the case, see ibid.
[28] Chris Albertyn, "Thor Executives Face Possible Murder Charges for Death of a Worker," reprinted in *Toxic Trade Update* 6, no. 3 (1993): 10; see also *Waste and Environment Today* 5, no. 4 (April 1992): 3–4, and 5, no. 5 (May 1992): 13.

continued to accept imported mercury wastes for which it was paid over U.S.$1,000 per ton. These wastes were improperly stored in warehouses on the plant site, waiting to be incinerated.[29] In 1997 a presidential commission of inquiry into Thor Chemicals deemed the mercury waste situation at the plant site to be "out of control."[30]

Environmental groups have also highlighted the problems associated with the trade in used plastics. Recycling of imported plastics has been found to have negative effects on the health of recycling plant workers and the environment in less industrialized countries. Asia has been a particularly popular destination for such wastes. Indonesia, for example, was the recipient of some 100,000–150,000 tons of plastic wastes from the United States and Germany every year in the early 1990s. According to Greenpeace reports, only 60 percent of this plastic was recyclable. The rest, a quarter of which was hazardous, was dumped into landfills. The plants that carried out the recycling of the remaining plastic operated under extremely unsafe conditions. Plastic wastes often contain residues from their previous contents, such as toxic cleaners, pesticides, and fertilizers. The recycling plants employed mainly women and children, many of whom did not wear protective clothing. The plastic waste import business has also been accused of destroying the livelihood of some 30,000–40,000 scavengers of local plastic waste (supporting some 200,000 dependents) who supplied small-scale plastic recycling firms.[31] These health, environmental, and social effects of imported plastic waste led the government to ban its import in 1993. But other countries still face the hazards of plastic waste imports, including China, the Philippines, Taiwan, India, Latvia and Nigeria, to name just a few.[32] Over 6 million kilograms of plastic wastes were shipped from the United States to the Philippines for recycling in 1991.[33]

Other wastes exported to developing countries for recycling have also been highlighted by environmental groups. The highly dangerous reclamation of lead from used lead-acid car batteries imported from industrialized

---

[29] Kockott 1994, 36.
[30] Arend Hoogervorst, "Mercury Waste Stockpile Riles South African Lawmakers," *Environmental News Service*, August 4, 1998, http://www.ban.org/ban_news/mercury.html; see also Earthlife Africa's website at http:www.earthlife.org.za.
[31] Ingo Bokerman and Jochen Vorfelder, *Plastics Waste to Indonesia: The Invasion of the Little Green Dots* (Hamburg: Greenpeace Germany, 1993).
[32] Greenpeace Canada, *We've Been Had! Montreal's Plastics Dumped Overseas* (Montreal: Greenpeace, 1993); "German Wastes Flood Latvia," *Toxic Trade Update* 6, no. 3 (1993): 11–13.
[33] Greenpeace, *The Waste Invasion of Asia* (Sydney, Australia: Greenpeace, 1994): 20–22.

countries is thriving business in Asia and Latin America. Used car batteries are exported to these regions from the United States, Australia, Japan, Canada, and the United Kingdom. These countries have an efficient system for collecting used batteries. But many lead recycling plants have been closed down in these countries because of the high cost of implementing environmental and health precautions for workers. The destinations of these used batteries are recycling operations in Brazil, Mexico, Indonesia, the Philippines, Thailand, and Taiwan, where standards are more lax and costs are much lower. Lack of protective clothing, unsanitary conditions, and poor ventilation are typical in these lead recycling firms. Most of the workers in these plants suffer from severe lead poisoning, and some have lost their lives as a result of lead contamination. Plant workers' families are also affected because lead residues are brought into the home on the clothes of the workers. The pollution emitted from these recycling plants also affects neighboring communities and wildlife through the contamination of the air, soil, and water from lead fumes and discharges.[34]

The Philippines imported some seventy-six thousand tons of lead-acid batteries from 1991 to 1996. The main sources of these batteries were Saudi Arabia, Singapore, and Australia, with some also coming from Canada, Japan, New Zealand, United Kingdom, Germany, and the United States.[35] Many of the batteries entering the Philippines from Singapore are en route from elsewhere, with Singapore acting as a transshipment center. For example, Philippine Recyclers Inc. (PRI), located near Manila, recycled twenty-four thousand tons of car batteries per year in the early 1990s.[36] PRI is an affiliate of the U.S.-based company Ramcar Batteries. Greenpeace studies at PRI have shown that levels of lead in and around this plant are extremely high and that elevated levels of lead in the blood of children and workers has led to a host of health problems.[37] Indonesia also has several lead-acid battery recycling plants and imports batteries for these from Australia and the United Kingdom in particular.[38] The largest of these plants, Indo Era Multa Logam (IMLI) located south of Surabaya, processes some sixty thousand tons of lead-acid batteries each year, almost

---

[34] Madeline Cobbing, *Lead, Astray: The Poisonous Lead Battery Waste Trade* (Amsterdam: Greenpeace International, 1994); Bill Moyers and CIR, *Global Dumping Ground* (Cambridge, U.K.: Lutterworth Press, 1991), 52–61.
[35] Greenpeace, "Philippines Fails to Halt Toxic Waste Imports," *International Toxics Investigator* 8, no. 4 (1996): 13.
[36] Charles Wallace, "Asians Vie for Toxic Trade as Danes Seek Ban on Waste from West," *Vancouver Sun*, March 24, 1994, p. A17.
[37] Greenpeace 1996, 12.
[38] Cobbing 1994, 4–5.

all of which are imported.[39] The Indonesian government has attempted to control these processors since the early 1990s, leading to the closure of several plants.

In addition to recycling, exports of expired pesticides and other chemicals to countries outside of the OECD have increased. Eastern Europe has been particularly affected by this practice. In the early 1990s, over one thousand tons of banned and expired pesticides as well as other obsolete chemicals laden with DDT and mercury were shipped to Romania by a German firm, under the label of "humanitarian assistance." Environmental groups revealed that this waste was improperly stored in corroded barrels stacked out in the open air. The barrels burst in the extremes of heat and cold. The chemicals contained in them leaked into the ground near an apple orchard and into streams that fed drinking water wells. This incident, along with others in Romania that involved toxic waste and outdated chemicals, prompted public demonstrations. In response, the government announced that it would ban such imports.[40] Germany was eventually forced to take the wastes back following pressure from the Romanian government and from Greenpeace, which brought several of the barrels back to the German border.[41] Following this incident, another German firm proceeded to send hundreds of tons of banned and obsolete chemicals and pesticides to state firms in Albania, again under the label of "humanitarian assistance." To dispose of these wastes properly in the country of manufacture would have cost some U.S.$3,000–7,000 per ton. When the importing firms in Albania discovered the nature of the imports, they requested that the shipments stop. The exporters responded by sending yet more shipments.[42]

Mozambique is also currently facing a controversy regarding a proposal by the Danish government development agency to incinerate outdated pesticides and other hazardous wastes that have been building up at various sites throughout the country. Most of these chemicals and wastes were originally imported. Environmental groups are seeking to halt this project on the grounds that incineration will produce toxic fumes. There are also

---

[39] Cobbing 1994, 9.
[40] Andreas Bernstorff and Katherine Totten, *Romania: Toxic Assault* (Hamburg: Greenpeace Germany, 1992).
[41] "Germany Agrees to Take Back Hazardous Wastes," *Managing Hazardous Wastes*, no. 3 (1993): 4; "Germany Finally Retrieves Its Wastes from Romania," *Toxic Trade Update* 6, no. 1 (1993): 15–16.
[42] "German Waste Traders Dump 'Humanitarian Aid' in Albania," *Toxic Trade Update* 6, no. 1 (1993): 16.

worries that such a plant might act as a magnet for illegal imports of toxic waste.[43]

Waste traders have also continued to disguise wastes by mixing them with other products. In one case in the early 1990s, one thousand tons of copper smelter furnace dust (containing high levels of lead and cadmium) from a copper recycling firm were mixed with fertilizer by several individuals and firms from the United States.[44] This fertilizer was subsequently sold to the government of Bangladesh with assistance from the Asian Development Bank. When the scandal was uncovered, the exporters were brought to court by the U.S. government. The firms and waste traders involved were convicted and fined U.S.$1 million. But while the trial was in progress, the contaminated fertilizer remained in Bangladesh for over a year. Although the Bangladesh government was able to hold three-quarters of the substance, technically toxic waste, in a warehouse, a quarter of the fertilizer was sold on markets several months after the scandal broke. Some Bangladeshi farmers had already spread it on their fields before they were made aware of its contents.[45] Some of the fine that was charged to the exporters was used to return the unused fertilizer to the United States, and to help mitigate damage done to the farmers in Bangladesh.[46]

Environmental NGOs and the Evolution of the Basel Ban

The political response by environmental NGOs to the continuation of waste exports for recycling was to push for a ban on the trade between rich and poor countries within the Basel Convention. The regional and unilateral hazardous waste import bans were important in that they sent a clear message to waste-exporting states that the poor world is off limits for the waste trade. But these bans effectively put the burden of identifying waste imports on the importing states, which do not always have the financial and technical resources to monitor and enforce import bans. It was soon recognized that if such bans were to be effective, they needed to be supported by a global ban in which both sending and receiving countries

---

[43] "Danish Incineration Project Could Make Mozambique African Toxic Dumping Ground Says New Action Group," Press release, Environmental Justice Networking Forum, Greenpeace, BAN, Multinational Resource Centre, August 11, 1998, http://www.ban.org_news/incinerate.html.
[44] *Waste and Environment Today* 5, no. 6 (1992): 16, and 5, no. 7 (1992): 18.
[45] "US Toxic Waste Sold as Fertilizer in Bangladesh," *Toxic Trade Update* 6, no. 1 (1993): 13.
[46] "US Waste Trade Companies Must Retrieve Waste Contaminated Fertilizer from Bangladesh," *Toxic Trade Update* 6, no. 4 (1993): 8–9.

would cooperate. Because the Basel Convention is the only global waste treaty, it was seen as the most promising path to implementing a successful ban. A key part of environmental NGOs' campaign for this global ban was to gather evidence of the environmental impact of both hazardous waste disposal and recycling in poorer countries.

The issue of a waste trade ban between industrialized and less industrialized countries within the context of the Basel Convention was debated at the first conference of parties to the convention (COP-1), held in Piriapolis, Uruguay, in late 1992. At the meeting, UNEP's then executive director, Mostafa Tolba proposed that a decision be adopted calling for a full ban on toxic waste exports from OECD to non-OECD countries, including waste exports destined for recycling operations. This proposal was endorsed by the G-77 countries present, which had, along with China, formalized their position at the UNCED preparations earlier that year in favor of a full ban on the hazardous waste trade with non-OECD countries. Environmental groups, Greenpeace in particular, also supported UNEP's proposal for a decision on a ban. This NGO came to COP-I well prepared to present extensive evidence on the environmental harm caused by both waste disposal and hazardous waste recycling operations.

Many countries and environmental groups at the meeting were in favor of this full ban on the waste trade with non-OECD countries. But the six largest producers of hazardous waste (United States, United Kingdom, Germany, Australia, Japan, and Canada) were firmly opposed. Part of the reason for this dissent was the language of the proposal, which divided countries into OECD and non-OECD groupings. Non-OECD would include Eastern and Central Europe, a region that some OECD countries wanted to keep open as an option for their waste exports. These six countries were also firmly opposed to any ban on waste exports destined for recycling operations. Several other OECD countries, however, did not agree with this position. Denmark, for example, surprised the rest of the EC by promising to use its EC presidency to press for an outright ban on both disposal and recycling from the EC to non-OECD countries. At the meeting, the G-77 group of developing countries, Poland, Hungary, Switzerland, Sweden, Norway, Finland, and Italy all agreed on the need for such a ban.[47] In addition to the six OECD countries that opposed the full ban,

---

[47] *ENDS Report*, no. 215 (December 1992): 37; "The Dirty Half-Dozen Stand Alone," *Toxic Trade Update* 6, no. 3 (1993): 2–4.

some Eastern European countries such as Russia and Estonia were not opposed to imports of waste for recycling.[48]

The Basel Convention secretariat and UNEP officials were keenly interested at that point in increasing membership in the Basel Convention among the largest waste-producing industrialized countries so as to give the treaty more legitimacy. It was thus worried that pushing too hard for a ban would scare away the key industrialized countries and ultimately weaken the impact of the treaty. Tolba noted at the meeting, "Industrialized countries generate more than 95 percent of the world's hazardous waste, and the USA is the biggest generator. Without their ratifications and active participation in implementing the treaty, obviously the Basel Convention will get nowhere."[49] The rules of the convention call for all steps to be taken to ensure that decisions are made by consensus rather than by a vote. Given the importance to the secretariat of keeping the participation of the largest waste producers, these countries were able to wield considerable influence even though they had not yet ratified the convention. The United States, Japan, the United Kingdom, and Germany threatened to hold back ratification if the convention went too far in their view.

Greenpeace campaigned heavily for a ban at this meeting. It lobbied both OECD and G-77 governments. It showed videos on the dangers of the waste trade in the breaks between sessions and handed out research papers on the effects of the waste trade and dirty recycling. Greenpeace was successful in its mission to disseminate important evidence on dumping and dirty recycling and in its goal to form a broad coalition among the G-77 and some OECD countries in favor of a ban. But in order to reach a consensus among the parties, a compromise had to be made with the opposing industrial countries. Following late-night negotiating sessions at which the United States and Canada argued strongly against a full ban, a compromise was finally reached. The decision adopted was not a firm ban but rather a "request" that "developed" and "developing" countries refrain from engaging in the waste trade with one another. The wording of this request was important because "developing" countries excluded Eastern and Central Europe, leaving those regions open to exports of waste from OECD countries. Moreover, the trade in hazardous waste destined for

---

[48] Debora MacKenzie, "'Business as Usual' for Traders in Toxic Waste," *New Scientist* 136 (December 12, 1992): 9.
[49] "Transfrontier Waste Meeting Focuses on Exports, Liability," *ENDS Report*, no. 215 (December 1992): 37.

recycling operations was explicitly exempted from this request.[50] This move to separate disposal from recycling was seen by environmental groups and many developing countries as a dangerous precedent.

It was agreed with the adoption of this request that the issue of a full and binding ban between OECD and non-OECD countries, including the waste trade for recycling, would be reconsidered at the next conference of parties in light of new evidence. The NGO–Third World alliance thus won a partial victory. Its cause dominated the meeting, and it was given the opportunity to present its case for a full ban again within three years. Indeed, Greenpeace was one of the more informed sources on the global trade in toxic wastes. Because it had the most ties to other NGOs involved in the campaign to end the waste trade, it was clear that it was up to Greenpeace to provide most of the proof of dirty recycling in developing countries at the second conference of parties to be held in 1994. After the first conference of parties, Greenpeace, in cooperation with local NGOs in developing countries, carried out further extensive research on the issue and prepared detailed reports for distribution. In that same period many developing countries ratified the Basel Convention with the encouragement of Greenpeace in order to increase the number of parties to the convention in support of a full ban on the hazardous waste trade with non-OECD countries.[51] This marked a change in Greenpeace's original stance that developing countries refrain from ratifying the convention until it was strengthened.

Although the parties to the Basel Convention were requested not to trade in wastes between developed and developing countries at COP-1, the waste trade with poor countries did not abate significantly following the meeting. Southeast Asia and Eastern and Central Europe became the new favorites with waste traders, particularly for waste exports destined for recycling operations. This was no doubt in part a response by waste exporters to the bad press they had received about exporting waste to the poorest countries. They hoped that sending waste for recycling to rapidly industrializing countries would not bring as much criticism.

When the second conference of parties to the Basel Convention (COP-2) was held in March 1994, the issue of whether to adopt a full ban on the waste trade between OECD and non-OECD countries again dominated the meeting. Sixty-four countries had become parties to the convention by

---

[50] Jim Vallette, "Basel 'Dumping' Convention Still Legalizes Toxic Terrorism," *Toxic Trade Update* 6, no. 1 (1993): 2–3.
[51] Interview with Jim Puckett, Greenpeace International, Amsterdam, October 1993; Interview with Kevin Stairs, Greenpeace International, London, November 1993.

this time. Most of these countries were present at the meeting, as were several NGOs, intergovernmental organizations, and industry representatives. Of the six industrial countries firmly opposed to the ban at the previous conference of parties, the United Kingdom, Japan, Canada and Australia were now full voting parties, while the United States and Germany (the latter as part of the EU) had yet to ratify the convention. Most of the less industrialized countries that had recently ratified the convention were present at the meeting, and they far outnumbered the more industrialized country parties.[52] Greenpeace was the most vocal NGO present and played a very active role in the plenary and working group discussions and behind the scenes negotiations, as well as advising the G-77 on strategy and wording of proposals.[53]

From the outset of the meeting there was general agreement on the need to ban hazardous waste exports destined for disposal between OECD and non-OECD countries. The parties, including the six countries that opposed the ban at COP-1, did not attempt to refute the overwhelming evidence presented by NGOs that such dumping was likely to be environmentally unsound. The meeting was extremely tense, however, as there was a deadlock over the issue of banning the trade in toxic wastes between OECD and non-OECD countries for recycling purposes. On the first day of the meeting, the G-77 (along with China) put forward a proposal calling for a ban, by mid-1996, on OECD exports of hazardous wastes to non-OECD countries for both recycling and disposal. In a separate proposal, Denmark also called for action to adopt such a ban. Its proposal was strongly supported by the other Nordic countries. The European Union put forward a third proposal that called for a ban on the trade for disposal between OECD and non-OECD countries but allowed for the continuation of the trade for recycling purposes for non-OECD countries that invited it. As the only document on the table in favor of a continuation of the waste trade with non-OECD countries for recycling purposes, the EU proposal threw the debate open. The endorsement by the EU of a continuation of waste exports to poor countries for recycling was seen by Greenpeace and others in favor of a ban as a serious obstacle to their goal. They

---

[52] The African parties that acceded to the Basel Convention in 1993 include Egypt, Mauritius, Seychelles, and Tanzania. Of the African parties to Basel, only Tanzania and Mauritius had also acceded to the Bamako Convention at that time.

[53] Greenpeace's position was outlined in the report "The Case for Prohibiting Hazardous Waste Exports Including Recycling from OECD to Non-OECD Countries," prepared for the second meeting of the Conference of Parties to the Basel Convention, Geneva, Switzerland, March 21–25, 1994.

were fearful that poorer countries could be bullied into accepting such imports if this proposal was adopted. Indeed, there were accusations at the meeting that certain OECD countries threatened to withdraw economic assistance to non-OECD countries if they did not alter their position regarding the recycling issue.

The G-77 countries, working in close consultation with Greenpeace, did not weaken under this pressure. They vowed to bring the matter to a vote if their demand for a total ban on waste exports from OECD to non-OECD countries was not met. If brought to a vote, it was clear that the ban decision would easily be approved by the required two-thirds majority. The countries opposed to the full ban attempted to weaken the coalition in favor of a ban by causing delays on action regarding the various proposals on the table. The threat by the United States not to ratify the convention did not have as much weight as it did at the first conference of parties because most other OECD countries had ratified the convention. But Canada and Australia, as full parties, attempted to delay any decisions by drawing out discussion in the working group meetings. They also tried to limit the activities of the NGOs present. Despite these tactics, the G-77 held more influence than it did at COP-1. It hung tightly together and was not willing to budge on the issue of a complete ban or on its position that NGOs should be allowed access to the working group meetings. The G-77 did make it clear, however, that it would consider a compromise on the date at which the ban on recycling would become effective. Although many of the EU country delegates supported of the total ban, they stated that they must adhere to any consensus reached within the EU if the matter came to a vote. Because the United Kingdom and Germany remained firmly opposed to a ban, it was not clear that a consensus in favor of a total ban would be reached within the EU over the course of the meeting.

In an attempt to reach a consensus, Greenpeace helped the G-77 draft a compromise position. This compromise proposal still insisted on the total ban but altered the phaseout date for a ban on recycling from June 30, 1996, to December 31, 1997. Still there was no agreement from the opposing countries. The G-77 kept open the option of calling for a vote if its demands were not met. But the key OECD countries let it be known that if they were seen to lose a vote, they would withhold funding to the secretariat of the Basel Convention to carry out the decision.[54] Just as the G-77 countries were at the point of calling for a vote on the issue to ensure that

---

[54] "Waste Exporters Lose Battle of Geneva—But the Fight over Scrap Metal Goes On," *ENDS Report*, no. 230 (March 1994): 16.

a decision was made at this meeting, word on the proposals came from the EU environment ministers who were meeting in Brussels at the time. They announced that the EU had agreed to a total ban on waste exports for recycling to non-OECD countries, as of January 1, 1998. This agreement came after the ministers from the Nordic countries intensely lobbied the United Kingdom and German environment ministers.

With the EU now in agreement, Canada, Australia, Japan, and the United States were left isolated as the only four countries against the full ban. As the United States was not yet party to the convention, it had little weight for bargaining, and the other three countries were resigned to going along with the consensus decision to ban the trade. On principle, the convention was not amended to include this ban. Rather, the decision that was adopted (Decision II/12) prohibited the trade with non-OECD countries on the grounds that such trade in wastes had "a high risk of not constituting an environmentally sound management of hazardous wastes as required by the Basel Convention."[55] Greenpeace agreed at the time that an amendment to the convention was not necessary and that the decision clarified the existing convention.[56] The U.S. delegation did not openly state that it would refuse to sign the convention now that this decision was taken, but it did hint that it might reconsider its position.

The adoption of the decision on the morning of the final day of COP-2 brought cheers from environmental groups and the delegates from the countries in favor of the ban. But it was immediately recognized that implementing the decision would be a difficult task. The decision called for individual states to report to the secretariat of the convention on its implementation, rather than active monitoring by the secretariat. The secretariat lacks the funding to monitor implementation of the decision, and there are no severe sanctions to ensure compliance. For these reasons it was left largely up to recipient states and environmental NGOs to make sure that the decision was adhered to by the waste-exporting states. The NGOs and developing countries themselves did not have the power to impose sanctions against those countries that did not comply, but they could pursue public campaigns to embarrass states that opposed the ban. The influence of environmental NGOs, and Greenpeace in particular, was a crucial force behind the adoption of the ban decision. Elizabeth Dowdswell, then newly appointed executive director of UNEP, at the close of the conference stated that environmental NGOs had become key

---

[55] UNEP/CHW.2/CRP.34 (adopted March 25, 1994).
[56] Interview with Kevin Stairs, Geneva, March 24, 1994.

Box 3.2 Text of Decision II/12

The Conference,

Recalling the request of the G-77 countries at the First Meeting of the Conference of the Parties to the Basel Convention in Uruguay, 30 November–4 December 1992, for the total ban on all exports of hazardous wastes from OECD countries to non-OECD countries;

Recognizing that transboundary movements of hazardous wastes from OECD to non-OECD States have a high risk of not constituting an environmentally sound management of hazardous wastes as required by the Basel Convention;

1. Decides to prohibit immediately all transboundary movements of hazardous wastes which are destined for final disposal from OECD to non-OECD States;

2. Decides also to phase out by 31 December 1997, and prohibit as of that date, all transboundary movements of hazardous wastes which are destined for recycling or recovery operations from OECD to non-OECD States;

3. Decides further that any non-OECD State, not possessing a national hazardous wastes import ban and which allows the import from OECD States of hazardous wastes for recycling or recovery operation until 31 December 1997, should inform the Secretariat of the Basel Convention that it would allow the import from an OECD State of hazardous wastes for recycling or recovery operations by specifying the categories of hazardous wastes which are acceptable for import; the quantities to be imported; the specific recycling/recovery process to be used; and the final destination/disposal of the residues which are derived from recycling/recovery operations;

4. Requests the Parties to report regularly to the Secretariat on the implementation of this decision, including details of the transboundary movements of hazardous wastes allowed under paragraph 3 above. Further requests the Secretariat to prepare a summary and to compile these reports for consideration by the Open-ended Ad Hoc Committee. After considering these reports, the Open-ended Ad Hoc Committee will submit a report based on the input provided by the Secretariat to the Conference of the Parties of the Convention;

5. Requests further the Parties to cooperate and work actively to ensure the effective implementation of this decision.

Source: Secretariat of the Basel Convention, *Report of the Second Meeting of the Conference of the Parties to the Basel Convention on the Control of Transboundary Movements of Hazardous Wastes and Their Disposal*, Geneva, March 21–25, 1994, UNEP/CHW.2/CRP.34.

players in post-UNCED environmental politics, and that the Basel decision was an example of their positive influence.[57]

The legal status of the decision, however, was the subject of a great deal of debate following COP-2. Some, especially those opposed to the ban, argued that it had no legal status because it was not an amendment to the convention. Others, including Greenpeace, argued that it clarified the existing convention and that an amendment was not required for the parties to be bound by the decision. Business interests were clearly upset with the decision. The U.S. Chamber of Commerce, for example, withdrew its support for the U.S. ratification of the Basel Convention less than two months after the ban decision was made.[58] U.S. movement on ratification ground to a near halt after the ban decision, and some have argued that withdrawal of support of the treaty by the U.S. Chamber of Commerce was the reason.[59] There was also some resistance to implementing Decision II/12 in the EU. But by April 1995 it did pass a modification of Regulation 259/93 on the Supervision and Control of Shipments of Waste Within, into and out of the European Community which effectively banned, as of January 1, 1998, all exports of hazardous wastes from the EU to non-OECD countries.[60]

Because of the confusion surrounding the legal status of the ban decision, Senegal suggested that a special workshop be held to discuss its implementation. With support of UNEP, a meeting was held in Dakar in March 1995 for that purpose. On the agenda was a discussion of how to clarify definitions of hazardous waste to ensure that all parties would be satisfied with the ban. The meeting was politically charged, with Greenpeace accusing the International Chamber of Commerce (ICC) of trying to sabotage the ban.[61] In its submission to the meeting, Greenpeace published a series of leaked documents, one of which indicated that the ICC's strategy for the meeting was to encourage non-OECD countries to oppose the ban. In return, the ICC planned to give funding to local industry groups in developing countries.[62] Harvey Alter, as a representative of the

---

[57] UNEP Information, press release, March 10, 1994.
[58] "U.S. Business Group Withdraws Support for Basel Treaty after Ban on Waste Trade," *International Environment Reporter* 17, no. 11 (June 1, 1994): 463.
[59] "U.S. Failure to Ratify Basel Treaty Seen as Environmental Justice Issue," *International Environment Reporter* 17, no. 22 (November 2, 1994): 891.
[60] "European Commission Approves Ban on Shipments of Waste for Recycling," *International Environment Reporter* 18, no. 9 (May 3, 1995): 317–18.
[61] "Meeting Participants Fail to Agree on Basel Treaty Definitions for Banned Waste," *International Environment Reporter* 18, no. 6 (March 22, 1995): 211–12. See also Greenpeace, *Implementing the Basel Ban: The Way Forward*, prepared for the Global Workshop on the Implementation of DecisionII/12, Dakar, Senegal, March 15–17, 1995.
[62] Greenpeace 1995. See especially leaked document 6, in Appendix 4 to this report.

ICC, did not deny this strategy, saying that the ICC was trying to get industries in South America, Southeast Asia, and Eastern Europe to oppose any amendment to the Basel Convention that incorporated the ban.[63] The documents also showed that the U.S. government planned to take active steps at the Dakar meeting to encourage non-OECD governments to reconsider the ban decision as well.[64] But the meeting did not have many concrete results. No major decisions were made, but there was agreement that further work needed to be done to clarify definitions of hazardous waste in the Basel Convention.

Following the Dakar meeting, environmental groups as well as many countries began to recognize that for the ban to be accepted and adhered to, the convention should be formally amended to incorporate the ban. This issue dominated COP-3, held in September 1995, in Geneva. For countries that favored the ban, as well as for environmental groups, the battle had to be fought and won yet again. But if they were successful this time, there would be no doubt as to the ban's legal status. Greenpeace's strategy at COP-3, as in previous COPs, was to villainize those countries opposed to the ban. It once again made use of moral arguments but this time made it clear that legitimate trade in scrap for recycling was not endangered by the ban. This move was an attempt to mitigate the efforts of the scrap recycling lobby, which had become fiercely interested in the question of the Basel ban following the ban decision taken at COP-2. Indeed, there was a sharp rise in the number of industry groups participating as observers at COP-3.

The Nordic countries put forward a proposal at COP-3 requesting a formal amendment to the convention in accordance with Decision II/12. This proposal was supported by most developing countries and by the EU, and Greenpeace. Predictably, industry groups and the main producers of hazardous waste were aligned in their opposition to the adoption of a ban amendment along with the United States, Canada, Australia, and Japan. At this meeting, however, for the first time, some non-OECD governments, including those of Brazil, India, Russia, and South Korea, also expressed opposition to the adoption of a ban amendment. Though these countries expressed concern, a consensus decision to adopt the amendment was easily reached. Several countries did register reservations after the amendment was adopted. Canada and Australia, for example, stated

---

[63] "Countries Developing List of Materials That Would Fall under Proposed Trade Ban," *International Environment Reporter* 8, no. 15 (July 26, 1995): 563–64.

[64] Greenpeace 1995. Again, see especially leaked document 6, in Appendix 4 to this report.

that they would not consider ratifying the amendment until certain definitional issues regarding what constitutes a hazardous waste had been worked out. Russia went further by stating that it could not accept the amendment.[65]

The amendment, adopted under Decision III/1, banned export of hazardous waste between OECD and non-OECD countries immediately and banned exports of hazardous waste destined for recycling as of January 1, 1998. This amendment will come into force once it is ratified by 62 of the parties that were present when it was adopted at COP-3. The wording of Decision III/1 was not identical to that of Decision II/12. The primary difference was that the categories of OECD and non-OECD were removed from the decision. Instead, it refers to parties listed in Annex VII and those not listed. Annex VII, which was also adopted at this time, lists parties and other states that are members of the OECD, EC and Liechtenstein. These countries are to refrain from the export of wastes to countries not listed in Annex VII, once the amendment is ratified.

It was unclear at the time that the Basel Ban Amendment was adopted what impact it would have on the status of Article 11 agreements, which allow for bilateral waste trade agreements between parties. Language making it clear that Article 11 agreements would be allowed under the ban was removed from the amendment at the last minute, leaving the question unresolved. It was also unclear whether developing countries could freely join Annex VII on their own initiative, which in effect would allow them to opt out of the ban at will. This uncertainty surrounding these two issues, as well as the vague definitions of hazardous wastes in the convention, angered industry representatives. They felt that it simply left too many loose ends. Environmental groups were firm at the time that there should be no exceptions to the ban under Article 11 and that countries should not be allowed to freely join Annex VII. At the same time that the ban amendment was adopted, however, the parties agreed to give the Technical Working Group (TWG) of the convention the task of drafting more precise definitions of what constitutes a hazardous waste and to clarify the Article 11 and Annex VII issues. This was seen as a measure to placate the growing international recycling industry lobby and those governments that had expressed reservations about the amendment.

---

[65] See UNEP, *Report of the Third Meeting of the Conference of Parties to the Basel Convention on the Control of Transboundary Movements of Hazardous Wastes and Their Disposal* (Annexes - I-III), UNEP/CHW.3/34, Geneva, September 18–22, 1995; see also "Ban on Waste Exports Outside OECD Pushed through Basel Treaty Meeting," *International Environment Reporter* 18, no. 20 (October 4, 1995): 753–54.

Box 3.3   Text of Decision III/1

The Conference,

Recalling that at the first meeting of the Conference of the Parties to the Basel Convention, a request was made for the prohibition of hazardous waste shipments from industrialized countries to developing countries;

Recalling decision II/12 of the Conference;

Noting that:

– the Technical Working Group is instructed by this Conference to continue its work on hazard characterization of wastes subject to the Basel Convention (decision III/12);

– the Technical Working Group has already commenced its work on the development of lists of wastes which are hazardous and wastes which are not subject to the Convention;

– those lists (document UNEP/CHW.3/Inf.4) already offer useful guidance but are not yet complete or fully accepted;

– the Technical Working Group will develop technical guidelines to assist any Party or State that has sovereign right to conclude agreements or arrangements including those under Article 11 concerning the transboundary movement of hazardous wastes.

1. Instructs the Technical Working Group to give full priority to completing the work on hazard characterization and the development of lists and technical guidelines in order to submit them for approval to the fourth meeting of the Conference of the Parties;

2. Decides that the Conference of the Parties shall make a decision on a list(s) at its fourth meeting;

3. Decides to adopt the following amendment to the Convention:

Insert new preambular paragraph 7 bis:

Recognizing that transboundary movements of hazardous wastes, especially to developing countries, have a high risk of not constituting an environmentally sound management of hazardous wastes as required by this Convention;

> Insert new Article 4A:
>
> 1. Each Party listed in Annex VII shall prohibit all transboundary movements of hazardous wastes which are destined for operations according to Annex IV A, to States not listed in Annex VII.
>
> 2. Each Party listed in Annex VII shall phase out by 31 December 1997, and prohibit as of that date, all transboundary movements of hazardous wastes under Article 1(i)(a) of the Convention which are destined for operations according to Annex IV B to States not listed in Annex VII. Such transboundary movement shall not be prohibited unless the wastes in question are characterized as hazardous under the Convention.
>
> Annex VII
>
> Parties and other States which are members of OECD, EC, Liechtenstein.
>
> *Source*: Secretariat of the Basel Convention, *Report of the Third Meeting of the Conference of Parties to the Basel Convention on the Control of Transboundary Movements of Hazardous Wastes and Their Disposal* UNEP/CHW.3/34, Geneva, September 18–22, 1995.

Conclusion

The developments in the waste trade following the adoption of the Basel Convention as well as regional waste trade treaties illustrate the unpredictable outcomes that can arise in a dynamic global economy. Despite agreement on the various conventions that should protect developing countries from waste exports from industrialized countries, the waste trade persisted in the early 1990s, albeit in a different form. One of the major problems was the rapid growth in the export of waste under the label of recycling and further use. This development can be seen largely as a response to new regulations at the international level. Although the Basel Convention discouraged exports from rich to poor countries of waste for disposal purposes, it allowed for the trade in wastes considered to be raw materials for recycling or further use. In addition, OECD and EU legislation encouraged the trade in wastes for recyclables. These new regulations created incentives for waste traders to shift their trade from that for disposal to that for recycling and further use. The dynamic nature of the global economy facilitated this shift, enabling the trade in wastes to develop into a global recycling business.

Environmental groups and developing countries were enraged by this unexpected development. But they did not give up their fight for a ban on the waste trade. In the years following the adoption of the Basel Convention, the alliance between environmental NGOs and developing country governments made its biggest impact on the global waste trade rules. They uncovered important data on waste trade schemes and pushed hard for further changes to the Basel Convention. The national and regional waste trade bans and the Basel Convention conference decisions were officially adopted by states. But nonstate actors were extremely important in the process of their adoption. Greenpeace provided much of the evidence of environmentally unsound toxic waste trade and dirty recycling in non-OECD countries. At the Basel conference of parties meetings it took an active role not only in distributing its research results but also in embarrassing those countries opposed to the ban in its comments in the plenary and working group sessions. Greenpeace also worked closely with the G-77 countries on strategy and the wording of compromise proposals, as well as lobbying governments that were sitting on the fence. Though some business lobbyists were also present at the initial conference of parties meetings and attempted to present the other side, the evidence presented by Greenpeace and its effective strategy for conveying that information overshadowed the attempts of the business groups to make their case.

4
Industry Players and Post-Basel Ban Amendment Politics

When the Basel Ban Amendment was adopted in September 1995, its supporters were jubilant. Following years of debate, the global community had finally outlawed the trade in wastes from rich to poor countries. Industrial lobby groups, particularly those representing the international recycling industry, were furious because the ban included not just hazardous waste destined for disposal but also hazardous wastes destined for recycling operations. After the Basel Ban Amendment was adopted, recycling industry organizations became much more engaged in the Basel debates in an effort to weaken the Basel ban. Their strategy was twofold. First, they took it upon themselves to convince parties to the convention that they had made a grave mistake and argued that they must not ratify the ban amendment. Second, they became very active not just in the public COP meetings but also in the less public Technical Working Group meetings where more precise definitions of hazardous waste were being hammered out. The TWG was also the forum in which questions regarding Annex VII membership and Article 11 agreements in relationship to the ban amendment were to be sorted out. Industry tried to ensure that definitions of waste in the context of the Basel Convention excluded those recyclable wastes that were most economically important to them and tried to open up opportunities to circumvent the ban by expanding of Annex VII membership and Article 11 agreements. The attempts by industry to weaken the ban through these measures can be seen as yet another aspect of the hazard transfer problem. When new international rules were agreed upon to regulate transboundary movements of hazards for recycling, efforts were made to stop them from being implemented.

Environmental groups, Greenpeace in particular, were somewhat less active on the waste issue immediately following the adoption of the ban amendment. They had achieved their main goal, the adoption of the Basel ban, and were rethinking their next steps. But it soon became apparent that continued pressure was needed to counter the attack on the ban by industry groups. A new environmental group, an offshoot of the Greenpeace Toxics Campaign, the Basel Action Network (BAN), was established in early 1998 and has since that time taken the lead role among environmental NGOs on waste trade issues. As industry organizations branched out to lobby at the global level, environmental NGOs began to lobby more at the national and local levels.

In this chapter I examine post–Basel Ban Amendment politics. In particular I look at the increased involvement of the global recycling industry in waste trade politics and the shift in strategy taken by environmental groups. The changing strategies of these nonstate actors help to illustrate the dynamic nature of the hazard transfer problem. There was a shift in the debate to the Technical Working Group, a forum that gave nonstate actors, particularly industry, much more scope for wielding influence. In taking a key role in driving the main debates in the Basel negotiations, nonstate actors have strengthened their roles in this process.

The Global Recycling Industry and the Basel Convention

The campaign by environmental groups against the international trade in wastes destined for recycling operations has been a thorn in the side of the global recycling industry. The adoption of the ban amendment appeared to be a fairly clear-cut case of industry's interests being ignored in a multilateral environmental agreement. Calls went out from the business community for a challenge to the World Trade Organization. Industry argued that the Basel Ban Amendment was inconsistent with GATT/WTO rules. Although it appears that industry was defeated by the Basel ban, a challenge to the WTO has yet to materialize. Does this signal that industry has accepted its apparent defeat by environmental groups and the majority of the parties to the Basel Convention? Or could it be that industry has found ways to ensure that its interests are met in the context of the Basel ban? I argue that the latter is the case, leading to yet another aspect of the hazard transfer problem.

Worth some U.S. $160 billion per year and employing some 1.5 million people around the world in the late 1990s, the global recycling industry has become increasingly organized and vocal in its attempt to defend itself.[1] Most of the global trade in recyclable materials is not hazardous, but a small portion of it is. Industry's concern regarding a potential ban on the trade in recyclable wastes has stemmed from fears that what it considers to be legitimate trade will be controlled as hazardous waste. The definition in the Basel Convention of what constitutes a hazardous waste is extremely vague, as has been emphasized by legal experts since it was first adopted.[2] The convention itself initially defined hazardous wastes according to its Annexes I and III, which list waste stream categories and hazardous characteristics, respectively, and these definitions applied to both wastes destined for disposal operations and wastes that can be recycled. Industry has argued that the definitions of hazardous waste under the convention were ambiguous and inconsistent with many national definitions of hazardous wastes, making it very unclear as to whether the trade in scrap metals and other recyclables is to be controlled by the rules of the convention.[3]

As the prospect of a ban on recyclable hazardous waste came closer to reality, the recycling industry's presence at the Basel meetings increased dramatically and the membership of recycling industry organizations grew. The possibility of the recycling trade being controlled by a global environmental treaty that would decide which wastes would be covered by a ban made global lobbying essential. There are several other reasons why

---

[1] http://www.bir.org/uk/index.htm.
[2] See, for example, Katharina Kummer, *International Management of Hazardous Wastes: The Basel Convention and Related Legal Rules* (Oxford: Clarendon Press, 1995); Jonathan Krueger, "When Is a Waste Not a Waste? The Evolution of the Basel Convention and the International Trade in Hazardous Wastes," paper presented at the International Studies Association, March 1998.
[3] Harvey Alter, "Industrial Recycling and the Basel Convention," *Resources, Conservation and Recycling* 19 (1997): 30.

a more global strategy was necessary. The United States, a major exporter of such wastes, has not yet ratified the convention, and getting the U.S. government to oppose the ban would not be particularly useful since it is not a party to the treaty. This has not stopped industry from lobbying the U.S. government to take the position that it would not ratify the treaty if the ban were adopted. But this threat did not seem to dissuade the other parties from endorsing the ban. In addition, the European Union countries have participated in the Basel Convention as a group, and attempts by the recycling industry to lobby individual governments have not been enough to influence the EU position. Further, it was important for the recycling industry to take its lobby efforts to the global level to counter global-level lobbying by environmental NGOs, which the industry felt was undermining its reputation for being "environmentally oriented." The recycling industry thus had to operate at the global level in an attempt to make its concerns heard.

The increased participation by the recycling industry at both the Basel COP meetings and the more specialized technical and legal meetings is evidenced by the growth in the number of business groups that appeared on the COP and TWG attendance lists starting in the early 1990s. Indeed, no recycling industry representatives attended the initial negotiations of the Basel Convention in 1989 whereas their numbers jumped to over 10 groups represented at COP-3 and COP-4. It was not until it became apparent that environmental NGOs and a growing number of parties to the convention sought to end the trade in wastes destined for recycling and recovery operations that the recycling industry became involved at the international level.[4] Greenpeace lost no time in attacking the recycling industry's stand on the issue, raising concerns over numerous cases of "sham recycling" that the group had uncovered.[5] Since then the scrap recycling industry has had steady attendance at these meetings.[6]

The Bureau of International Recycling (BIR), based in Brussels, is the largest international recycling lobby group and has taken on the issue of the Basel ban with full force. The BIR's members are some six hundred firms and national recycling federations in over fifty countries. The recycling activities of its members include ferrous and nonferrous metals, textiles,

---

[4] Peggy Abrahamson, "Basel Treaty Is Getting a Green Light," *American Metal Market* 99, no. 202 (October 21, 1991): 7.

[5] See, for example, Jim Puckett, "Disposing of the Waste Trade: Closing the Recycling Loophole," *Ecologist* 24, no. 2 (1994): 53–58.

[6] Michael Marley, "BIR, Greenpeace Haggle over Toxics: Defining Hazardous Waste Key Issue," *American Metal Market* 100, no. 143 (July 24, 1992): 9.

paper, plastic, rubber, and glass. It claims that its ultimate aim is environmental preservation and argues that a global free market in recyclable materials is the best way to achieve this goal. Other industry groups that are concerned about the recycling issue and have been actively lobbying at the global level include the International Council on Metals and the Environment (ICME), the International Precious Metals Institute (IPMI), the Institute of Scrap Recycling Industries (ISRI), the Business Recycling Council (BRC), and the umbrella industry lobby group, the International Chamber of Commerce. Harvey Alter, head of the BRC, estimates that some five hundred individuals with various industry lobby groups are in close contact with one another regarding Basel issues.[7]

Initially some members of the recycling industry were supportive of the Basel Convention and wanted the U.S. government to ratify it not only so it could influence the direction of the treaty but also so that the United States could more easily participate in trade with other parties. At least one industry representative, John Bullock, legal counsel for a U.S.-based scrap metal firm, urged the precious metals recycling industry at an event sponsored by the ISRI to pay more attention to the Basel Convention and argued for U.S. ratification.[8] But the recycling industry's support for the treaty was tempered by the growing demands by developing countries and environmental groups in the early 1990s for incorporation of a ban on the waste trade in the treaty. The marked increase in the number of industry representatives present at COP-2 was in part because it was known that a motion would be put forward at the meeting for the parties to the convention to adopt a decision with respect to a ban on the waste trade in the context of the treaty. John Bullock, who attended the COP-2 as a representative of the ICC, carried around a computer circuit board during the entire meeting. He used it to demonstrate to delegates that the export to developing countries of computer waste, which has a high content of valuable scrap metal, might be outlawed if a ban was adopted. He argued that such a ban would harm economic prospects for developing countries that relied on scrap metal imports as an affordable way to obtain precious metals.

Despite industry's lobbying at the meeting, a ban on the waste trade between rich and poor countries was adopted by the parties in a consensus decision. But in late 1994, Greenpeace and the BIR met to discuss the ban issue. The business press reported that a representative from Greenpeace

[7] Interview with Harvey Alter, Washington, D.C., May 19, 1999.
[8] Lynne Cohn, "Recyclers Are Asked to Push for Basel Treaty; Handy and Harman Executive Says It Is Broader than US Law," *American Metal Market* 101, no. 240 (December 14, 1993): 7.

noted that the areas of agreement between the two groups were greater than the areas of disagreement and that the environmental group really only objected to about 5 percent of the waste traded for recycling by BIR members.[9] But when the European Commission announced its plan to ban waste exports for both disposal and recycling to non-OECD states according to the COP-2 decision, the BIR expressed its annoyance. The BIR pledged at that time to direct its political lobbying efforts toward influencing the definition of wastes at the international level.[10]

The recycling industry was dealt a further blow when it was unable to stop the momentum toward adopting the ban amendment at COP-3 in September 1995. Industry members noted that the 1995 meeting had good attendance by industry representatives from both developed and developing countries and was "less of a 'green party' meeting than some of the previous meetings."[11] Industry representatives made it clear that they wanted a delay in any decision regarding the trade in recyclables until the definitional issues in the Basel Convention were solved. They were frustrated when the Basel Ban Amendment was adopted by consensus at COP-3 in 1995 because they wanted the issue brought to a vote. The head of the BIR, Francis Veys, said at the time: "It was disgusting, the president forced a consensus; many people were not happy with it."[12]

Industry's Arguments against the Basel Ban

The recycling industry, especially exporters of scrap metal, clearly have a lot at stake. According to an UNCTAD study, the share of scrap metal exports from OECD countries that went to non-OECD countries jumped from 5.2 percent in 1980 to 29 percent in 1993. The same report notes that about a third of OECD trade in scrap and metal residues makes its way to non-OECD countries, most of going to developing countries in Asia.[13] Trade in

---

[9] Gloria LaRue, "Recycler-Environmentalist Détente Takes Uneasy Road," *American Metal Market* 103, no. 104 (May 31, 1995): 5A.
[10] Ibid.
[11] More than sixty industry observers, including twenty-two from developing counties, were present. Michael Marley, "Hazardous Waste Battle Lingers," *American Metal Market* 103, no. 190 (October 3, 1995): 6.
[12] "An End to 'Toxic Colonialism'?" *Chemistry and Industry*, no. 19 (October 2, 1995): 759.
[13] Cited in OECD, *Trade Measures in the Basel Convention on the Control of Transboundary Movements of Hazardous Wastes and Their Disposal*, COM/ENV/TD(97)41/FINAL (OECD: Paris, 1998), 24.

secondary metals is worth about U.S.$14 billion annually, and the value of trade between Annex VII countries and non-Annex VII countries is about U.S.$4 billion per year.[14] And net exports of waste from the United States alone that could be affected by the trade ban (excluding iron) are estimated to be U.S.$ 2.5 billion per year.[15]

The recycling industry quickly launched a global campaign against the Basel Ban Amendment. A key part of its strategy was to claim that recyclable scrap should not be considered as waste. Scott Horne of the ISRI said that the problem with the Basel Ban can be summed up in four words: "Scrap is not waste."[16] Industry pressed this message in its efforts to try to convince countries that had not yet ratified the agreement to refrain from doing so. The BIR included an extensive section on its website dedicated to arguments against the Basel ban.[17] The ICME responded by commissioning a study on the trade policy implications of the Basel Ban Amendment. This report recommended that parties not ratify the amendment, or at the very least not until the uncertainties such as the definition of hazardous waste, the status of Article 11 agreements, and the rules for joining Annex VII were cleared up. It argued that the last resort would be to raise a challenge to the WTO.[18]

Industry also used its campaign against ratification of the Basel Ban Amendment to portray itself as a victim in the process and as a vehicle to express its views regarding what it sees as the positive relationship between free trade and environmental protection. The main problems with the ban as put forward by industry are that it will be difficult to implement, that it will produce undesirable outcomes, and that it risks jeopardizing other important achievements.[19]

The first argument by the recycling industry against the Basel ban is that it will not work. Industry has consistently argued that a ban on the trade in recyclable hazardous waste between rich and poor countries

---

[14] Philip Burgert, "Metal Recyclers Face Definition Dilemma," *American Metal Market* 105, no. 45 (March 6, 1997): 5.
[15] http://www.bir.org.
[16] Interview with Scott Horne, Washington, D.C., May 19, 1999.
[17] The BIR has posted its views on the Basel Ban Amendment on its website. Access to its newsletter is limited to members who pay a fee to join. http://www.bir.org.
[18] Maria Isolda Guevara and Michael Hart, *Trade Policy Implications of the Basel Convention Export Ban on Recyclables from Developed to Developing Countries* (Ottawa: ICME, 1996), vi.
[19] The industry's arguments fall neatly into the three categories outlined by Albert Hirschman as being common rhetorical positions taken by the right in reaction to progressive changes to existing rules. These are arguments based on futility, perversity, and jeopardy. See Albert O. Hirschman, *The Rhetoric of Reaction: Perversity, Futility, Jeopardy* (Cambridge, Mass.: Harvard University Press, 1991).

would be impossible to implement without further refinement of the definition of hazardous waste. The main fear has been that with vague definitions in the convention, a ban might stop trade in recyclables that the industry does not consider to be hazardous or even as waste. With numerous definitions floating around and each country having its own definition, the chances that wastes that are not considered hazardous by the exporting country will be banned according to the convention or refused because the importing country does consider such waste to be hazardous seemed very high. The BIR felt that imposing a ban before specifying what is and is not a hazardous waste would make it impossible to implement fairly. The ISRI in early 1997 adopted a position statement opposing U.S. ratification of the Basel Convention if no steps were taken to clarify its definitions of hazardous waste. It was hopeful, though, that some resolution could be reached. ISRI's director of government affairs, Thomas Wolfe, said of the need to improve definitions of waste in the text of the treaty, "It won't make us love the Basel Convention, but it will make us hate it less."[20] As the work of the TWG progressed, it was clear that these arguments were no longer the strongest ones in its arsenal so industry began to focus more on other arguments.

A second argument made by industry groups against the Basel ban is that the proposed changes will have the opposite effect than that intended. John Bullock, for example, claims that the ban is based on "inverted logic" because it will not result in the most environmentally sound outcomes.[21] First, it is argued that it will discourage recycling and result in more use of virgin materials, which, according to the recycling industry, will ultimately cause more damage to the natural environment. This position has been promoted consistently by Rafe Pomerance, U.S. deputy assistant secretary for environment and development (and, interestingly, a former head of Friends of the Earth International).[22] Industry groups also claimed that if local recycling industries in the poorer countries are denied enough stock to make them efficient and profitable, they may fold and then recyclable materials may end up in final disposal. Again, this is a suboptimal outcome for the environment.[23]

---

[20] "ISRI Adopts Resolute Stance against Basel," *American Metal Market* 105, no. 1 (January 2, 1997): 7.

[21] Cited in "Summary of Panel Discussion, Trade and Environment: Challenges for 1996," Session #3: New Development in the Basel Convention, Summary Report, January 19, 1996. http://www.gets.org/gets/library/admin/up_d_the_Environment_Challenges_for_1996_htm.

[22] Cited in "An End to Toxic Colonialism?" *Chemistry and Industry*, no. 19 (October 2, 1995): 759; and Rob Edwards, "Leaks Expose Plan to Sabotage Waste Treaty," *New Scientist*, 145 (February 18, 1995): 4.

[23] Guevara and Hart 1996, vi.

A further perverse outcome stressed by industry is that the adoption of the ban could seriously harm the economic prospects of developing countries because recyclable scrap metal is much less expensive for developing countries to obtain than virgin materials.[24] Banning the trade in scrap is seen to "run counter to elemental economic logic."[25] It is argued that the ban in such trade would be a severe economic blow to developing countries and would hinder their chances of achieving "sustainable development." As Harvey Alter, a staunch opponent of the Basel ban argues:

> The majority of "hazardous wastes" moving in intentional trade are for recycling and are important sources of secondary raw materials and energy conservation for developing countries. Any restriction on trade threatens to eliminate a traditional and major source of raw materials, threatens to undermine the industrial growth that underpins economic development (which funds advances in environmental protection and management) and is contrary to the principles of sustainable development.

Industry also warns that the ban will likely lead to an increase in illegal traffic. Bullock warns the international community that the already existing problem of illegal trade in wastes will only increase if the trade is banned outright and that this will lead to undesirable situations that will subject workers and the environment to unnecessary harm.[26]

The third argument made by industry is that commonly held core values will be jeopardized if such a ban is implemented. In particular, it argues that the Basel Ban Amendment puts the global free trade system at risk. Industry firmly believes that growth is essential to the attainment of sustainable development globally and that threats to free trade are a direct attack on growth. The ICME was so annoyed by this perceived contradiction of free trade principles that it argued that a challenge to the WTO over this issue should be considered.[27] A study commissioned by ICME conveys the sense of risk, stating, "Ultimately, the challenge lies in meeting both domestic and environmental objectives without undoing the benefits of an open, multilateral trade regime."[28]

---

[24] "Q: Why Is the Scrap Metal Trade Important to Developing Countries?" *ICME Newsletter* 6, no. 2 (1998): 7.
[25] Rod Hunter, "Good Intentions, Foolish Policy," *Chemistry and Industry*, no. 2 (January 15, 1996): 68.
[26] Cited in "Summary of Panel Discussion" 1996.
[27] Guevara and Hart 1996, vii.
[28] Ibid., vii.

The sovereign rights of developing countries are also seen by industry to be put at risk by the ban. Environmental groups argued that the sovereignty of developing countries is enhanced by the ban because they collectively agreed that they did not want to receive hazardous wastes and the ban would respect these wishes. But industry has argued the opposite, that governments of developed countries are being patronizing and effectively telling developing countries what they can and cannot do.[29] According to industry groups, the ban denies developing countries their right to exercise any comparative advantage they may have in recycling.[30] Indeed, Francis Veys commented at COP-3 that "some of the delegates I spoke with from Africa and Asia said they see this as invasion of their national sovereignty. They feel that the green lobby, particularly the Europeans, are trying to impose rules on them."[31] The ICC has referred to the ban as a form of "eco-imperialism ... whereby industrialized countries impose their environmental standards on developing countries."[32] Taking the issue further, the study commissioned by the ICME lamented that excluding developing countries from receiving recyclables is a violation of their GATT/WTO rights.[33]

Finally, the recycling industry has argued that the ban on the trade in recyclables would seriously jeopardize jobs and investments, especially in developing countries. The BIR, for example, has complained that investments in developing countries in environmentally sound technologies are at risk and that shutdowns of recycling operations in developing countries will result in thousands of job losses.[34]

It is unclear whether the recycling industry will succeed in dissuading parties from ratifying the ban amendment. Twenty-four ratifications had been deposited with the Basel secretariat as of April 2001, though sixty-two ratifications are needed to bring the amendment into force.[35]

---

[29] See, for example, Alter 1997, 49.
[30] Hunter 1996, 68.
[31] Michael Marley, "Hazardous Waste Battle Lingers," *American Metal Market* 103, no. 190 (October 3, 1995): 6.
[32] Cited in "Export Ban is Seen as Economic Blow," *American Metal Market* 105, no. 71 (April 14, 1997): 7.
[33] Guevara and Hart 1996, iv.
[34] http://www.bir.org.
[35] The following had ratified the ban amendment as of April 17, 2001: Andorra, Austria, Bulgaria, Cyprus, Czech Republic, Denmark, Ecuador, Finland, Gambia, Luxembourg, Netherlands, Norway, Panama, Paraguay, Portugal, Slovakia, Spain, Sri Lanka, Sweden, Trinidad, Tobago, Tunisia, United Kingdom, Uruguay, and The European Community.

Environmental NGO Strategies in Post-Basel Ban Amendment Politics

While industry strategy to influence Basel politics began to include global-level lobbying against the ban, environmental groups also began to shift gears. The initial strategy of environmental NGOs working on the toxic waste trade had been to encourage the adoption of a ban on the waste trade between rich and poor countries in the context of the Basel Convention. Once that goal had been achieved, first with Decision II/12, and then with the Basel Ban Amendment (Decision III/1), the environmental groups began to focus more on convincing individual countries to ratify and implement the ban amendment. They also moved toward a more reactive mode, responding to industry efforts against the ban. Greenpeace continued as the lead environmental NGO on the issue immediately following the adoption of the ban. It vowed to continue to monitor the countries and industries that might attempt to undermine the ban. It soon became apparent that the adoption of the ban amendment, though not yet in force legally, had the effect of significantly reducing the number of attempted waste exports to developing countries. In response to this, Greenpeace soon scaled back much of its staff time in the campaign against the toxic trade. This organization began to reduce Basel-related activities in its national offices, though Greenpeace International's Political Unit continued to attend the principal Basel meetings. The organization's toxic trade campaign newsletter was renamed the *International Toxics Investigator* in 1996. It covered news on toxic technologies and persistent organic pollutants as well as the waste trade. This publication was phased out completely in 1997, presumably with the understanding that most of this information now could be posted on its website.

As virtually the sole NGO operator in the Basel fight at the international level up to that point, Greenpeace had been very effective. Because it was the principal coordinator of environmental groups opposed to the waste trade and the main organization attending the Basel negotiating sessions, it was able to stand firm on its principles and did not have to make many compromises on its strategy. It was able to dictate the terms of the NGO debate. But after the Basel Ban Amendment was adopted, Greenpeace was not seen by those working on the issue to be the best suited organization to achieve the new goals of ratification and implementation of the Basel Ban Amendment at the national level. It had national offices in only about thirty countries, which was not considered sufficient to drum up national-level support in key countries to secure ratification of the

Basel Ban Amendment. This was also a time when membership and financial support for Greenpeace had been dropping in many countries. As industry's campaign heated up against the ratification of the ban in various countries, it became clear that more broad-based grassroots support in favor of the ban ratification was needed. Within Greenpeace, it was recognized that other national-level NGOs, which had until then been relatively less involved in the issue or had simply followed Greenpeace's lead, needed to be brought more fully into a broad-based coalition. This coalition needed to be based on a more equal agenda-setting among organizations because environmental groups in each country had to follow a unique strategy to convince their governments to ratify the ban.[36]

The coordinator of Greenpeace's international toxic trade campaign, Jim Puckett, felt that it was important to shore up Greenpeace's political and technical expertise with a large grass-roots network of locally based NGOs to persuade their national governments to ratify the Basel Ban Amendment. Following this vision, he launched a new group (BAN), in 1997.[37] By early 1999 BAN already had 23 member organizations in 15 countries in both the industrialized and developing worlds.[38] This network still maintains close ties to Greenpeace in coordinating strategy, but is focused primarily on the ratification and implementation of the Basel ban in as many countries as possible at as early a date as possible. Its strategy to achieve this goal was very different from its strategy in calling for the ban to be adopted. To convince parties to adopt the ban amendment it was necessary to show that there were numerous cases of waste trade between rich and poor countries. But to convince them to keep and ratify the ban amendment, it became necessary to demonstrate that the ban, which although it was not yet in force, had the effect of reducing waste exports to developing countries, and was a success in averting a major environmental crisis. This was a much harder task. Though the strategy of the environmental NGOs involved in the toxic trade issue has shifted to include more grassroots and national-level activities in support of the Basel Ban Amendment, its global-level strategy is still very strong. Representatives from the

---

[36] Jim Puckett, director of BAN, personal communication, July 26, 1999.
[37] BAN, "Global Toxics Network Launched to Safeguard Toxic Waste Export Ban," press release, http://www.ban.org/ban_news/launched.html.
[38] Helge Ole Bergesen, Georg Parmann, and Oystein Thommessen (editors), *Yearbook of International Co-operation on Environment and Development 1999/2000* (London: Earthscan, 1999), 260. The countries where BAN-affiliated members are located are: Brazil, Bulgaria, Canada, Croatia, France, Haiti, India, Indonesia, Malaysia, Netherlands, Norway, Solomon Islands, South Africa, Taiwan, and the United States.

BAN secretariat are regular participants at the Basel meetings, and the group endeavors to bring member organizations to these meetings as well.[39] Greenpeace has continued to attend the Basel meetings and maintains its toxics campaign in coordination with BAN.

Defining Hazardous Wastes: The Basel Technical Working Group

The recycling industry and environmental groups have been very active in the Technical Working Group (TWG) of the Basel Convention, which was mandated at COP-3 to define hazardous waste more clearly. Industry groups saw the formation of this group as an opportunity to influence the convention. A representative from the BIR was quoted as saying that although they lost the battle of the ban, the war was far from over.[40] A principal goal of the recycling industry was to ensure that most of the scrap items traded internationally were not defined as hazardous waste by the convention. Francis Veys, head of the BIR, has stressed that the definition of waste must be dealt with as "a matter of urgency" at the global level rather just accepting national definitions.[41] Greenpeace and BAN were also very active in the TWG meetings. Greenpeace had initially claimed that only the definitions of scrap metals and residues required further elaboration.[42]

At the four meetings of the TWG between December 1995 and February 1997 the main agenda item was the definition of hazardous waste. To clarify which wastes are covered by the convention, the TWG began working on developing A, B, C and D lists of wastes according to their hazard characteristics. List A includes those wastes for which trade between Annex VII and non-Annex VII parties is banned. List B includes wastes for which trade between these groups of countries is allowed (unless determined to be hazardous by importing, exporting or transit countries). List C includes wastes on which decisions are pending. Wastes on List D arouse environmental concerns but are outside the scope of the convention. The president of COP-3, Bakary Kante, has described this system of classifying wastes as a "scientific approach."[43] The ultimate aim of the

---

[39] http://www.ban.org.
[40] Marley, "Hazardous Waste Battle Lingers," 6.
[41] Cited in Burgert 1997, 5.
[42] Greenpeace International, *The Basel Ban: The Pride of the Basel Convention* (Amsterdam: Greenpeace International, September 18, 1995), 4.
[43] Bakary Kante, "The Basel Convention: Promoting Environmentally Sound Management," *ICME Newsletter* 6, no. 2 (1998): 3.

TWG meetings was to finalize lists that could be adopted by the COP-4 held in Kuching, Malaysia, in early 1998, finally clarifying which wastes are to be covered by the convention.

Attendance at the TWG meetings ballooned following the adoption of the Basel Ban Amendment. A quick glance at the attendance lists for these meetings shows the heavy presence of industry groups. While no more than one or at most two environmental groups attended these meetings (primarily Greenpeace and at one meeting a local environmental NGO), there were between seven and twelve industry groups present, their numbers increasing with each successive meeting. The industry delegations were large; one NGO observer noted that at the September 1996 TWG meeting some 159 delegates were present (including both governmental and observer representatives) of which 49 were from industrial organizations.[44] Industry groups present at the TWG meetings would caucus daily to discuss their strategies.[45] Nonstate actors had a greater chance to influence the convention here than working through governments at the more public COP meetings, especially because some of the scientific and technical information the group required to complete its work could be provided only by industry. Because its representatives far outnumbered environmental groups, industry had significant opportunities to influence discussions on the definition of waste.

The global recycling industry's goal in participating in the TWG was to ensure that the B list included the vast majority of wastes that it was already exporting and that it be formally adopted by the parties at the COP-4. Industry wanted to ensure that the B list of wastes was based on "sound science." It was also concerned that some wastes were "mirror listed," that is, they appeared on both the A and B lists. Industry argued that this needed to be corrected, preferably by putting these wastes on the B list. Here the issue was that some wastes are listed as "hazardous" if they are contaminated with hazardous substances, but they may not exhibit hazardous characteristics. The BIR wanted assurance that such shipments would not be face delays in order to be certified as "nonhazardous." Industry also wanted to ensure that the status of Article 11 agreements would be clarified by the TWG. Finally, it wanted clarification of procedures for countries that are not currently listed in Annex VII to add their names to that list, which would enable them to opt out of the ban.[46]

---

[44] Jim Puckett, "The Basel Ban: A Triumph over Business as Usual" October 1, 1997, http://www.ban.org/about_basel_ban/jims_article.html.

[45] Interview with Harvey Alter, Washington, D.C., May 19, 1999.

[46] These positions are outlined in the BIR website: http://www.bir.org.

The BIR was involved directly in the classification process in the TWG from early on and has been somewhat optimistic about the outcomes. Veys of the BIR remarked, "We are hopeful these lists based on science, technology and fact will enable the non-OECD countries to receive the materials they need."[47] The BIR in fact commissioned a science professor from the United Kingdom to research and provide scientific data on whether certain tradable scrap materials are hazardous, and this research was presented to the TWG. BIR has claimed that it "has been instrumental in securing B list classification for the majority of secondary materials traded by its members."[48] But at the same time, industry has been cautious not to appear to be too influential. At the first TWG meeting in late 1995, a representative from the ICC made a statement which stressed that industry wanted to "avoid any false impression that within the Basel Convention, industry will be writing legal texts." The ICC representative also stressed that industry wanted to ensure that its work was reviewed by member states.[49]

Environmental groups' goals at these meetings were primarily to ensure that industry groups did not have undue influence in defining wastes. They critiqued industry submissions, and pressed for designation on List A of those potentially hazardous wastes which industry claimed were best put on List B. Some contentious issues did arise, though, as NGOs were accused of making personal criticisms of some delegates.[50] The TWG finished its classification work in early 1997. For the most part, the process was judged by both industry and environmental groups to have been fairly legitimate.[51]

In addition to its lobbying efforts to influence the Basel Convention in the latter half of the 1990s, the global recycling industry also launched a concerted attempt to redefine all recyclable material as "recycled raw materials." In other words, it aimed to eliminate the use of the word "waste" in relationship to recycling.[52] The BIR vowed in 1996 to make a

---

[47] Quoted in Michael Marley, "Basel Pact Conferees Affirm Bans on Scrap," *American Metal Market* 103, no. 185 (September 26, 1995): 1.
[48] http://www.bir.org.
[49] Secretariat of the Basel Convention, *Report of the Technical Working Group of the Basel Convention*, Ninth Session, Bonn, December 11–13, 1995, paragraph 28. Interestingly, environmental NGOs were not mentioned.
[50] Secretariat of the Basel Convention, *Report of the Technical Working Group of the Basel Convention*, Thirteenth Session, Geneva, April 27–29, 1998.
[51] See, for example, Bette Hileman, "Treaty Grows Less Contentious," *Chemical and Engineering News*, April 6, 1998, 29; Puckett 1997, 12.
[52] Tsukasa Furukawa, "BIR Battling 'Waste' Label: Trade Group Leader Promises Two-pronged Attack," *American Metal Market* 104, no. 106 (May 31, 1996): 7.

"two-pronged attack" by pushing for this new label through both technical means (via the Basel TWG) and legal means (by raising court challenges).[53] Following lobbying by the recycling industry, in 1997 the U.S. Environmental Protection Agency decided to exclude certain scrap metals, including "unprocessed production and home scrap, processed scrap and agglomerated fines and drosses" from its definition of solid waste.[54] In an attempt to spur similar definitional changes in other countries, the BIR and some of its national member organizations have brought forward legal challenges on definitions of waste in the United Kingdom, the EU, and Australia, calling for a reclassification of recyclable material as a "nonwaste."[55] In the United Kingdom, for example, the British Metal Federation (with support of the BIR, which has set up a legal fund to launch court cases regarding waste definitions) challenged the government's definition of waste, which includes secondary metals. The BIR sees this as an important strategy that should be carried out in European countries and could advance its cause of redefining scrap metals as "products" rather than wastes.[56] This court challenge was eventually struck down. The EU warned that it would not follow the United States by removing scrap from its definition of waste.[57] By the end of the 1990s, EU recycling firms finally dropped their campaign to remove scrap from the definition of waste. Instead, they began to push for the EU to produce information sheets on types of scrap in order to clarify what is and is not a waste on a case-by-case basis.[58]

Debates over the Basel Ban at COP-4 and COP-5

The work of the TWG was presented at COP-4, which was held in Kuching, Malaysia, in February 1998.[59] This meeting was attended by several NGOs associated with BAN, including Greenpeace, as well as by a roughly equal number of industry groups. One of the more important decisions slated for adoption at COP-4 was that associated with two new annexes to the convention, Annex VIII and Annex IX, which represent the A

---

[53] Ibid.
[54] http://www.bir.org/uk/keyissue.htm.
[55] Puckett 1997, 13.
[56] Krueger 1998, 17–18.
[57] Camilla Reed, "Interview—Scrap Metal Needs Waste Tag—E.U. Official," *Reuters*, March 2, 1999, http://www.ban.org/ban_news/interview.html.
[58] "E.U. Firms Reluctantly Accept Waste Definition," *ENDS Daily*, February 23, 1999.
[59] Secretariat of the Basel Convention, *Report of the Fourth Meeting of the Conference of Parties to the Basel Convention*, UNEP/CHW.4/35, Kuching, Malaysia, March 18, 1998.

list wastes and B list wastes respectively. This clarification of A and B list wastes and their codification in the convention was viewed by industry as one of the most important obstacles that needed to be overcome following the adoption of the Basel ban. Since the recycling industry itself claims that it has had an important influence on the process of developing those lists, one can assume that these lists met most of industry's needs.

The COP also requested the TWG to continue its work and in particular to provide the COP with recommendations for the revision and/or adjustment of the A and B lists in the future. This has left open the possibility that some wastes may move from one list to the other or that new wastes may be added. Such an occurrence could work both for and against industry's interests, depending on whether new wastes are added to the A or to the B list. As industry has continued to participate in the TWG meetings, it is presumed that it will continue to try to ensure that such a procedure for adjusting the lists is fair in its estimation.

While the adoption of annexes for classification of wastes was one of the more important decisions taken at the COP, it was not the most contentious. The question of Annex VII membership sparked the most debate at the meeting. Annex VII issues were raised because Israel, Monaco, and Slovenia had applied to join Annex VII. There were no rules or procedures set out on how to deal with these applications, and they became the topic of much discussion and debate at COP-4. Both BAN and Greenpeace argued strongly that freely allowing countries to join Annex VII would completely undermine the ban amendment because countries could willingly opt out of it. Further, these groups argued that it was pointless to change an amendment that had not yet come into force. This view was supported by the African countries, the Arab League, and several other developing countries. On the other side, some OECD countries and some developing countries such as Chile, Brazil, South Africa, and Argentina argued for the development of criteria that would lay out just how new members could be added. Canada took the most extreme position, arguing that any country that wished to join Annex VII should be freely allowed to do so. The EU argued that only Monaco should be allowed to join because it was a special case.

A heated debate ensued during the meeting, and at one point a small "contact group" of parties, from which NGOs were excluded, was created to discuss the matter.[60] Finally, after long debates and confusion over whether a consensus had been achieved on the issue, the COP decided not to change

---

[60] BAN, "Basel Ban Victory at COP4," http://www.ban.org/isues_for_cop4/what_happened.html.

the countries listed in Annex VII until the ban amendment comes into force. As part of the same decision, it asked the TWG, in cooperation with the Sub-Group of Legal and Technical Experts, to undertake a study to "provide parties with a detailed and documented analysis that would highlight issues related to Annex VII."[61] This decision no doubt annoyed the recycling industry, which had been hoping for a clarification on the procedure by which countries could join Annex VII and thus opt out of the ban. BAN declared victory because it meant that attempts to undermine the ban by amending it would be held off until the amendment came into force.

The status of Article 11 was debated in the run-up to COP-4. Shortly after COP-3, the EU publicly stated that it interpreted the ban to mean that Article 11 agreements are not allowed.[62] Industry groups did not accept this opinion, and funded several studies on this question. These studies concluded that not allowing Article 11 agreements under the ban amendment could be interpreted as contradictory to the rules of the WTO.[63] BAN and Greenpeace argued that it was not legal to use Article 11 agreements to circumvent the ban.[64] COP-3 had designated the task of developing guidelines for Article 11 agreements to the TWG, but these were not completed before COP-4. The parties at COP-4 asked the TWG to draft a guidance document on the interpretation of the ban with respect to Article 11 in collaboration with the Consultative Sub-Group of Legal and Technical Experts.[65] This document was not due to be completed before COP-5.[66] Environmental groups were pleased at this move. Again, industry was disappointed by the lack of clarity on this issue. The European Commission began developing a common position regarding the relationship between Article 11 and the Basel Ban Amendment.[67] The

---

[61] Secretariat of the Basel Convention 1998, Decision VI/8.
[62] The full text of this legal opinion is on the BAN website: www.ban.org.
[63] See Guevara and Hart 1996; James Crawford and Philippe Sands, *Article 11 Agreements under the Basel Convention* (Ottawa: ICME, 1997).
[64] BAN, "Annotated Agenda and Summary Recommendations for the Fourth Conference of Parties to the Basel Convention," February 23–27, 1998, 7; Greenpeace, "COP4: The Issues at a Glance," http://www.greenpeace.org/toxics.html.
[65] Secretariat of the Basel Convention, *Report of the Fourth Meeting of the Conference of Parties to the Basel Convention*, UNEP/CHW.4/35, Kuching, Malaysia, March 18, 1998, Decision VI/1.
[66] Secretariat of the Basel Convention, "Draft Guidance Elements for Bilateral, Multilateral or Regional Agreements or Arrangements," note, Fourth Session of the Open-ended Ad Hoc Committee for the Implementation of the Basel Convention, UNEP/CHW/C.1/4/15, June 21–25, 1999.
[67] BAN, *Report on the 4th Open Ended Ad Hoc Meeting for the Implementation of the Basel Convention*, Geneva, June 21–25, 1999, 3.

lack of clarity on both Article 11 agreements and Annex VII membership sparked the ICME to release another report on the issue in an attempt to convince governments not to ratify the ban amendment. It claims that the Basel ban is inconsistent with WTO rules, and the continuing lack of clarity on Article 11 and Annex VII only makes the potential disruption to trade more pronounced.[68]

Following COP-4, industry lobby groups in the United States softened their stance against U.S. ratification of the convention but qualified this position by insisting that the United States ratify only the original convention without the ban amendment. The BRC indicated that it would like to see the United States adopt the 1989 convention, plus Annexes VIII and IX.[69]

COP-5, held in December 1999 in Basel, was a significant meeting, as it represented the tenth anniversary of the adoption of the Basel Convention. The issue of the ban had been the focal point of the first four COP meetings, but it was not as prominent on the agenda of COP-5. But industry and environment groups had as strong a presence as ever at this meeting. Some countries expressed concern at the meeting regarding the slow pace at which ratifications of the Basel Ban Amendment were being deposited with the secretariat. Nonetheless, several countries announced that they were well on the way to ratifying it.[70] This was in addition to the seventeen ratifications that had already been registered. According to BAN, opponents of the ban have more or less accepted that it will eventually come into force. But at the same time, it was clear that industry and certain OECD countries still aimed ultimately to circumvent the ban. Calls were made by some OECD countries for the development of criteria for joining Annex VII. This prospect worried BAN, which has repeatedly pointed out that the development of such criteria was not included in the mandate for the study regarding Annex VII.[71] Israel and Slovenia both resubmitted their requests to the secretariat that they be added to Annex VII at COP-5. According to BAN, these countries might be simply trying to show a track record of

---

[68] See Maria Isolda Guevara, "Basel Convention Ban Amendment: Arguments against Ratification," *ICME Newsletter* 7, no. 1 (1999): 5, 8. Full report: http://206.191.21/210/icme/baselted.htm.

[69] "Industry Groups Say They Would Support Basel Legislation under Certain Conditions," *International Environment Reporter* 21, no. 12 (June 10, 1998): 567–68.

[70] Secretariat of the Basel Convention, *Report of the Fifth Meeting of the Conference of Parties to the Basel Convention*, UNEP/CHW.5/29, December 10, 1999, Geneva, 9.

[71] BAN, *BAN Report and Analysis of the Fifth Conference of the Parties to the Basel Convention*, December 1999. On BAN website: www.ban.org.

consistent attempts to join Annex VII in case they choose to challenge the Basel ban at the WTO in the future.[72]

The status of other potential avenues for circumventing the ban, Article 11 agreements, and review procedures for determining whether wastes were assigned to Annex VIII and Annex IX did not raise much excitement at the meeting. The parties at COP-5 adopted a decision to extend the mandate of the TWG and the Consultative Sub-Group of Legal and Technical Experts to develop guidance documents on Article 11 agreements. The parties also adopted provisional review procedures for dealing with applications for amending Annexes VIII and IX.

The issue of the ban and ways to work around it were strong in the background to COP-5, but new issues dominated the meeting. The adoption of a protocol on liability and compensation was the main agenda item. The meeting also introduced for the first time the question of how the convention should deal with the issue of shipbreaking—the dismantling of old sips that have toxic components. Also slated for the meeting was the adoption of a declaration on environmentally sound management of wastes. Each of these new issues has important implications for the integrity of the Basel ban.

Between COP-4 and COP-5 efforts were stepped up to complete a protocol on liability and compensation. Work on this protocol had been ongoing since 1993. The drafters of the original convention had hoped to include provisions on liability and compensation for damages arising from the transboundary movement of wastes in the original convention, but agreement on these matters was not secured at that time. Instead, Article 12 of the Basel Convention calls on parties to develop such a protocol as soon as practicable. The impetus for this protocol has been concerns by developing countries who worried that they would not be able to afford to clean up unwanted waste dumps or spills of imported hazardous wastes. Negotiations on the Protocol on Liability and Compensation continued until the last moment to ensure its adoption at COP-5.[73] BAN and other environmental groups immediately criticized the protocol, claiming that it is not only weak but is actually counterproductive.[74] A more in-depth analysis of this protocol will be provided in Chapter 7.

---

[72] BAN, *Report on the 4th Open Ended Ad Hoc Meeting*, 1999.

[73] Daniel Pruzin, "Compromise Text on Basel Liability Protocol Completed; Final Draft Expected This Fall," *International Environment Reporter* 22, no. 14 (July 7, 1999): 570. See also Secretariat of the Basel Convention, *Report of the Ad Hoc Working Group of Legal and Technical Experts to Consider and Develop a Draft Protocol on Liability and Compensation for Damage Resulting from Transboundary Movements of Hazardous Wastes and Their Disposal*, Ninth Session, Geneva, April 19–23, 1999, UNEP/CHW.1/WG.1/9/2, April 28, 1999.

[74] BAN, *BAN Report and Analysis of the Fifth Conference of Parties*, 1999.

The issue of shipbreaking was addressed by the Basel Convention for the first time at COP-5. The environmental and health consequences of the dismantling of ships that are at the end of their useful life have been a growing concern of environmental groups such as Greenpeace and BAN. When old ships are decommissioned and taken apart, primarily to recover steel, there are toxic components that must be disposed of, including asbestos, PCBs, and toxic metals such as mercury and lead. Toxic wastes, though only around 5 percent of the total weight of an average ship, can be a significant source of hazards for workers and environmental pollution. This is particularly true in shipbreaking yards in developing countries, where precautions are much less strict than in similar operations in industrialized countries. Over the past few decades, the majority of shipbreaking has shifted from countries such as Great Britain, Taiwan, Spain, Mexico, and Brazil to poorer developing countries, particularly those in Asia, where few protective measures are taken.[75] Since 1998 there have been numerous reports in the media and by environmental groups regarding worker and environmental safety issues at shipbreaking yards, with particular attention paid to one of the most environmentally harmful sites, the Alang Shipyard in India.[76]

Shipbreaking first appeared on the Basel Convention agenda at a TWG meeting in April 1999, mainly in response to pressure from environmental groups.[77] Basel parties and environmental groups acknowledged that ships destined for breaking in other countries are considered wastes under the Basel Convention. This means that the export of ships destined for breaking is subject to prior notification where applicable. Such ships also cannot be exported if there is reason to doubt that they will be disposed of in an environmentally sound manner. And, according to Decisions II/12 and III/1, Annex VII parties are not allowed to export ships for breaking to non–Annex VII countries, which includes developing countries. Yet environmental groups have been worried that a legal loophole exists regarding ships destined for breaking. It is conceivable that a ship might be sold by an Annex VII country owner to a new owner in a non–Annex VII country but might not be labeled as bound for decommissioning until it is already in the non–Annex VII country. Indeed, with the vast majority of shipbreaking now taking place in developing countries (many of which are

---

[75] Judit Kanthak, Andreas Bernstorff, and Nitiyanand Jayaraman, *Ships for Scrap: Steel and Toxic Wastes for Asia* (Hamburg: Greenpeace, 1999), 6.
[76] BAN, "Shipbreaking and Its Response in India," April 1999.
[77] BAN, *Report on the Basel Convention 15th Technical Working Group and 2nd Consultative Sub-Group of Legal and Technical Experts*, April 26, 1999.

parties to the Basel Convention) and a significant portion of those ships being imported from industrialized countries (many of which are also Basel parties), it is clear that the Basel provisions are being flouted in some cases. The parties asked the TWG, in collaboration with the International Maritime Organization, to develop guidelines on the environmentally sound dismantling of ships.[78] The TWG was also asked to work with the Legal Working Group in studying legal aspects of ship dismantling under the Basel Convention. These reports are to be presented at COP-6. To date there has not been much intervention on the part of industry against such action, and the decision to undertake these studies passed without controversy.

Perhaps the most significant outcome of COP-5 was the adoption of a Declaration and Decision on Environmentally Sound Management of Hazardous Wastes.[79] The declaration sets out priorities for the second decade of the convention. It emphasizes the importance of prevention and minimization, as well as the environmentally sound management of recycling and disposal of hazardous wastes. This marks a new direction for the convention, as it broadens the focus of the convention beyond transboundary waste movements to encompass waste management within countries. A more detailed discussion of this declaration and its significance for preventing hazard transfer in the future will be presented in Chapter 7.

Conclusion

The efforts to weaken the Basel ban on the part of industry and key OECD countries that backed industry's position represent another aspect of the hazard transfer problem. When the ban was adopted, the attempts to circumvent it were immediate and multipronged. Industry's strategy involved not just public lobbying in which it argued that the Basel Ban Amendment should not be ratified by parties to the convention but also attempts to weaken the legal force of the ban through the more obscure Technical Working Group of the convention. These attempts included industry's involvement in the process of defining which wastes are and are

---

[78] Secretariat of the Basel Convention, *Report of the Fifth Meeting of the Conference of Parties to the Basel Convention*, UNEP/CHW.5/29, December 10, 1999, Geneva, Decision V/28. See also Secretariat of the Basel Convention, *Report of the Technical Working Group, Fifteenth Session*, April 28, 1999, UNEP/CHW/WG.4/15/12, 12–13.

[79] Secretariat of the Basel Convention, *Report of the Fifth Meeting of the Conference of Parties to the Basel Convention*, UNEP/CHW.5/29, December 10, 1999, Geneva, Decision V/33.

not covered by the convention, the discussions on whether countries can freely join Annex VII, and guidelines on whether Article 11 agreements with non-Annex VII countries would be allowed under the ban.

Industry has made concerted efforts in these areas, but it is not clear that the ban has been undermined yet. Though twenty-four parties have ratified the ban amendment so far, announcements were made at COP-5 which indicate more ratifications will soon be forthcoming. And while industry has definitely put its imprint on the process of defining wastes, it appears that environmental groups were also involved and are satisfied that the process was fair. Industry now appears to be focusing its attack on the ban on Annex VII and Article 11 interpretations. These issues have yet to be sorted out. Environmental groups are actively trying to discourage the outcomes on these issues that industry is pushing for.

The literature on nonstate actors has focused primarily on their role in attempting to lobby states regarding their positions in the negotiation of agreements and at COP meetings. Much less attention has been paid thus far to the technical side of the interpretation and implementation of these agreements, in which industry groups in particular are also active. Though the activities of industry on these technical matters are less visible than its rhetorical public lobbying campaigns, its influence in these settings is extremely important in determining the implementation and effectiveness of global environmental agreements. Indeed, the role of nonstate actors has become more secure in the process following the adoption of the Basel Ban Amendment in 1995. The Basel secretariat has noted that the role of nonstate actors has been important in the COPs and TWGs regarding scientific and technical information about the environmentally sound management of hazardous wastes.[80] Confirming this role, parties adopted a decision at COP-5 requesting the secretariat to continue its close partnership with both industry and environmental NGOs.[81]

---

[80] Secretariat of the Basel Convention, *Cooperation with U.N. Bodies and Regional Systems and Organizations, and Others* UNEP/CHW.4/21, October 1997.
[81] Secretariat of the Basel Convention, *Report of the Fifth Meeting of the Conference of Parties to the Basel Convention*, UNEP/CHW.5/29, December 10, 1999, Geneva, Decision V/13.

5
Foreign Direct Investment in Hazardous Industries

Hazard transfer from rich to poor countries is not confined to the export of hazardous wastes. A further dimension to the problem is investment by TNCs in hazardous industries in developing countries. This practice was first identified in the 1970s and is still a problem. At first debates emerged over whether this practice was driven by differentials in environmental costs and regulations. Most economists examining the question concluded that highly polluting TNCs were not relocating to the developing world to avoid environmental costs. This to them has become "common wisdom." Environmental NGOs, however, have highlighted cases of hazardous industry migration in their campaigns as yet another avenue for hazard transfer. With the adoption of the Basel Ban Amendment, environmental groups fear that hazardous industry migration will only accelerate unless measures are taken to ensure the adoption of cleaner production methods in both rich and poor countries.

In this chapter I argue that foreign direct investment in the most hazardous industries in the developing world does not fit neatly into the established common wisdom that dirty firms do not migrate for environmental reasons. There are changing circumstances that give reason to revisit the debate over industry location and the environment, especially with respect to the most highly hazardous industries. These include changes in the global economy and in both global and national environmental regulation. Whether or not TNCs in hazardous industries are investing in poor countries for environmental reasons alone, they are increasingly doing so and their impact on the environment has been negative in many cases.

TNCs and Hazardous Industry Location

TNCs tend to invest in industries that have on average a high environmental impact, making inquiry into their environmental performance particularly important.[1] Highly polluting and hazardous industries, including resource extraction, electronics, textiles, and heavy manufacturing, have strong TNC involvement. Though this pattern of foreign direct investment in highly polluting industries has affected both developed and developing countries, the latter are receiving a growing share of such investment, just as it is declining in importance in the former.[2]

This tendency of TNCs to invest in highly polluting industries and the growing presence of this type of investment in developing countries raises concern. The rate of industrial growth in hazardous industries in developing countries is greater than the rate of overall industrial growth in those countries. It is also greater than the rate of growth of hazardous industries in developed countries. This trend began in the 1970s, just as environmental regulations became more stringent in OECD countries.[3] Is this

---

[1] UN Transnational Corporations and Management Division, Department of Economic and Social Development, *World Investment Report, 1992* (New York: United Nations, 1992), 226.
[2] Ibid., 231; Patrick Low, "The International Location of Polluting Industries and the Harmonization of Environmental Standards," in *Difficult Liaison: Trade and the Environment in the Americas*, ed. Heraldo Muñoz and Robin Rosenberg (London: Transaction, 1993), 25.
[3] Robert Lucas, David Wheeler, and Hemamala Hettige, "Economic Development, Environmental Regulation, and the International Migration of Toxic Industrial Pollution 1960–88," *Policy Research Working Papers*, WPS 1602 (Washington, D.C.: World Bank, 1992), 14.

growth in toxicity of production a result of foreign direct investment? One of the central debates in the literature on TNC investment and the environment has been over whether highly polluting industries migrate to countries that have less strict environmental regulations. This debate surrounds questions of whether there is "industrial flight" of firms from countries that erect higher environmental regulations and whether firms seek "pollution havens" in countries with lower environmental standards.[4] Most studies focused on pollution abatement costs of firms and concluded that polluting firms as a whole do not relocate in developing countries to avoid the high cost of complying with environmental regulations at home.[5] According to some analysts, one of the key reasons is that compared to other costs, such as labor, pollution control costs in dirty industries account for a low percentage of overall value of output. It is often argued that pollution control costs rarely constitute more than 2 percent of sales, making it unlikely that this would be a sufficient reason to migrate.[6]

But several of these studies noted that there did appear to be evidence that some the *most hazardous* industries a clear, though limited, trend of relocation for environmental reasons has emerged.[7] H. Jeffrey Leonard

---

[4] For an excellent overview of the debate, see Peter Thompson and Laura A. Strohm, "Trade and Environmental Quality: A Review of the Evidence," *Journal of Environment and Development* 5, no. 4 (1996): 363–88. For specific studies on this topic, see Ingo Walter, "Environmentally Induced Industrial Location to Developing Countries," in *Environment and Trade*, ed. Seymour Rubin and Thomas Graham (London: Frances Pinter, 1982); H. Jeffrey Leonard, *Pollution and the Struggle for the World Product* (Cambridge: Cambridge University Press, 1988); Patrick Low and Alexander Yeats, "Do 'Dirty' Industries Migrate?" (World Bank Discussion Paper 159), in *International Trade and the Environment*, ed. Low (Washington, D.C.: World Bank, 1992); Charles Pearson, "Environmental Standards, Industrial Relocation, and Pollution Havens," in *Multinational Corporations, the Environment, and the Third World*, ed. Pearson (Durham, N.C.: Duke University Press, 1987); Nancy Birdsall and David Wheeler, "Trade Policy and Industrial Pollution in Latin America: Where Are the Pollution Havens?" in Low 1992; Barry Castleman, "The Export of Hazardous Factories to Developing Nations," *International Journal of Health Services* 19, no. 4 (1979): 569–606; UNCTC, *Environmental Aspects of the Activities of Transnational Corporations* (New York: United Nations, 1985).

[5] For example, Walter 1982; Leonard 1988; Pearson 1987. These studies were also based primarily on data for U.S. firms, and thus it is difficult to transfer these findings to other regions, such as Asia, without further empirical study.

[6] Low 1993, 28; Robert Repetto, *Trade and Sustainable Development*, UNEP Environment and Trade Series, no. 1 (Geneva: UNEP, 1994), 22–3; Michael Ferrantino, "International Trade, Environmental Quality and Public Policy," *The World Economy* 20, no. 1 (1997): 52.

[7] UNCTC, *Transnational Corporations in World Development* (New York: United Nations, 1988), 230; UN TNC and Management Division 1992, 223: Leonard 1988, 111; Walter 1982, 101; Castleman 1979. See also Barry Castleman, "Double Standards in Industrial Hazards," in *The Export of Hazard*, ed. Jane Ives (London: Routledge, 1985); and Castleman, "Workplace Health Standards and Multinational Corporations in Developing Countries," in Pearson 1987.

shows that some industries involved in the use and production of heavy metals, asbestos-containing products, leather tanning, benzedine dyes, copper smelting, and hazardous chemicals and pesticides, did relocate to countries with more lax regulations.[8] Pollution control costs in the United States in these highly hazardous industries are generally higher than for other polluting industries. For example, for 1991 they ranged from 2 to 3 percent of sales for the chemicals industry and as much as 6 percent of sales for the metal plating industry.[9] Relocation of hazardous industries also seemed to have occurred in special cases where certain conditions were present in industrialized countries such as instances of protests against the establishment of these firms, in industries that faced significantly high environmental costs, and in industries that faced declining demand.[10] The identification of this trend was based on a growing number of empirical case studies of specific firms that have relocated hazardous processes, though no aggregate figures were given. For example, there has already been widespread movement of extremely hazardous industry from Japan to poorer countries resulting largely from the Japanese public's concern over the environmental effects of those industries.[11]

Most of the studies that confirmed this tendency of some of the most hazardous industries to relocate for environmental reasons dismiss it as insignificant compared to trends for industry as a whole. The argument that polluting industry on the whole does not migrate for environmental reasons has thus become common wisdom. It is argued that the growing toxicity of industry in developing countries is a result primarily of their stage of industrial development rather than foreign direct investment per se. Some, such as Muthukumara Mani and David Wheeler, acknowledge that there may have been some displacement of pollution-intensive industry which seems consistent with the pollution haven hypothesis when they

---

[8] Leonard 1988.
[9] Robert Atkinson, "International Differences in Environmental Compliance Costs and United States Manufacturing Competitiveness," *International Environmental Affairs* 8, no. 2 (1996): 214–15.
[10] Walter 1982, 89–90; Leonard 1988, 111–14.
[11] Hans Maull, "Japan's Global Environmental Policies," *Pacific Review* 4, no. 3 (1991): 254–62; Walter 1982, 95; UNCTC 1985, 45. In the case of Japan, it is not clear whether this industry migration is limited to only the most hazardous of polluting industries. Between 1980 and 1983, some two-thirds to four-fifths of all pollution-intensive foreign direct investment from Japan was directed toward developing countries in Asia and Latin America. See UNCTC 1985, 38. For a more recent investigation of Japanese foreign direct investment in pollution-intensive industries, see Derek Hall, "Dying Geese: Japan and the International Political Ecology of Southeast Asia," paper presented at the International Studies Association annual meeting, Toronto, March 1997.

note that "pollution intensive output as a percentage of total manufacturing has fallen consistently in the OECD and risen steadily in the developing world. Moreover, the periods of rapid increase in net exports of pollution intensive products from developing countries coincided with periods of rapid increase in the cost of pollution abatement in the OECD economies."[12] But they then go on to argue that the effects have not had major significance. One reason they give for this is that the proportion of products from dirty industry which are exported has not risen significantly (i.e., there has been a rise in domestic consumption of pollution-intensive industry products). Another is that relocation may have been owing more to the presence of energy subsidies in developing countries at a time when energy prices were high in the OECD. And finally, they argue that as these economies experience economic growth, demand for basic industrial production levels off, and as incomes rise these countries begin to implement more stringent environmental regulations. Thus the pollution haven phenomenon, though it exists in some sectors, is seen to be merely transitory and not significant.[13]

Though some argue that the relocation of the most hazardous firms to developing countries is not a significant trend, the fact remains that it has occurred in the past. Moreover, recent changes in the global economic and regulatory context raise new concerns that this practice may become even more pronounced in the future. In light of this new context, there are two serious weaknesses with the earlier studies that argue that it is common wisdom that firms do not relocate for environmental reasons. Both of these weaknesses have resulted in an underestimation of firms' environmental costs and thus an underestimation of the "push" factors facing firms, particularly those in the most hazardous sectors.

First, recent changes in both local and international regulations regarding hazardous industries and their wastes have significantly raised environmental costs for firms in the most hazardous industries. The earlier studies on the question of firm migration were based largely on U.S. data from the late 1970s and early 1980s.[14] This was before the costs of disposal and treatment of extremely hazardous wastes skyrocketed in industrialized countries. Starting in the mid- to late 1980s environmental regulations,

---

[12] Muthukumara Mani and David Wheeler, "In Search of Pollution Havens? Dirty Industry in the World Economy, 1960–1993," *Journal of Environment and Development* 7, no. 3 (1998): 241.
[13] Ibid., 244.
[14] Only U.S. data were used in most of these studies. Some have noted that these data may not reflect conditions in other countries. See, for example, Ferrantino 1997, 45.

particularly those for hazardous waste in the United States, became much more ambitious.[15] Pollution control costs for the most hazardous firms most likely rose in tandem with these more stringent regulations. In the 1980s, however, many firms began to export these wastes to other countries at a fraction of the cost of disposal at home. This is particularly true for U.S. and European firms that in many cases were able to send their hazardous wastes to developing countries. The prohibition of the export of hazardous wastes from OECD to non-OECD countries via the Basel Ban Amendment may result in an even further rise in pollution abatement costs for firms in OECD countries. This will likely be the case for those firms that had previously exported their hazardous waste to developing countries. This development could increase the push factor, resulting in even more relocations of hazardous industrial processes to developing countries.

A second weakness of most of the studies that claim that migration of dirty industry is not likely to occur is that they measured environmental costs by looking only at the amount spent on pollution control or abatement, the key area of environmental expenditures made by firms in the 1970s.[16] This narrow interpretation of environmental costs does not capture all of the expenditures associated with firms' environmental behavior today. Rather, it focuses almost exclusively on the cost of installing scrubbers in smokestacks and hazardous waste dumping fees and often excludes other extremely important costs.

These other environmental costs are many. They include the cost of cultivating a "green" image at home through the establishment of environment divisions and publication of public relations literature on environmental practice. Perhaps more important, they also include liability and insurance costs in case of environmental damage, costs for protecting workers from environmental hazards, and delays while environmental impact assessments are carried out.[17] They also include forgone investments due to the costs of meeting more stringent regulations.[18] These broader environmental costs undoubtedly are much higher than the direct costs of pollution abatement alone, especially in the most hazardous industries for

---

[15] Richard Stewart, "Environmental Regulation and International Competitiveness," *Yale Law Journal*, no. 102 (1993): 2084–85.

[16] UNCTC 1985, 34; Repetto 1994, 7; Low 1993, 24.

[17] Bernard Fleet and Jeremy Warford, "Managing Hazardous Industrial Wastes in Developing Countries," *Ecodecision* 14 (Fall 1994): 42–44; Chapman et al., "International Law, Industrial Location, and Pollution," *Global Legal Studies Journal* 3, no. 1 (1995): 26–27.

[18] Stewart 1993, 2084.

which these other environmental costs are highest. A World Resources Institute study has shown, for example, that environmental costs of producing an agricultural pesticide at a DuPont plant were more than 19 percent of total manufacturing costs.[19] Failure to include these wider costs in studies on whether firms relocate in order to escape high environmental costs at home may indeed have understated the environmental cost differentials between highly industrialized and developing countries.

Thus the increase in pollution abatement costs in recent years, plus the rise in other environmental costs, resulted in significantly higher overall environmental costs, especially for firms in the most hazardous industries—costs that were not accounted for in earlier studies on industry location. These rising costs may well be stronger push factors for hazardous firms than was earlier presumed.

At the same time that hazardous industries' environmental costs have been rising in wealthier countries, the environmental costs associated with these industries in the poorer countries have remained low. It is well known that environmental regulations in poor countries tend to be relatively weak. The lax enforcement of those regulations that do exist in most developing countries is widely acknowledged as lowering environmental costs. In addition, liability concerns for TNCs, which are significant in industrialized countries, appear to be less prominent in developing countries. Charles Hadlock has pointed out that TNCs do not worry about liability or public image problems nearly so much in their foreign affiliates in developing countries as they do at home.[20] Environmental impact assessments are not always required, nor are they always carried out according to instruction, so that delays are usually not significant.

The impact of lower environmental costs in developing countries that are associated with lax regulation and enforcement is important to highlight because most of the studies on environment and industry location have focused almost entirely on how environmental cost and standards differentials affect industrialized countries, rather than their impact in developing countries. This weakness in the mainstream literature has been highlighted by Gareth Porter, who argues that environmental standards and cost differentials between rich and poor countries do not create a "race to the bottom" in industrialized countries so much as they create a "stuck

---

[19] Cited in World Resources Institute, *WRI Annual Report, 1995* (Washington, D.C.: WRI, 1995), 14. This report did not indicate what percentage of total sales of this product these environmental costs constituted.

[20] Charles Hadlock, "Multinational Corporations and the Transfer of Environmental Technology to Developing Countries," *International Environmental Affairs* 6, no. 2 (1994): 170.

at the bottom" problem in developing countries. He argues that this is particularly true for rapidly industrializing countries.[21] The reason, according to Porter, is that the competition among developing countries for foreign investment from abroad in certain sectors is much more fierce than in industrialized countries because there are a number of countries seeking to attract investment in similar lines of production. The problem is compounded by the fact that governments in developing countries are much less responsive to their citizens' demands for stricter environmental regulations as well as enforcement of those regulations. While it is difficult to prove conclusively with hard data that lowering environmental standards reduces firms' costs, Porter argues that the widely held *perception* among government officials and investors that stricter environmental regulations will mean higher costs and reduced competitiveness does drive these actors to behave as if they do.[22]

Without downplaying the significance of weak domestic political institutions in explaining lax environmental regulations in developing countries, it is also important to highlight the role of global economic factors. In the 1980s and 1990s many developing country governments found themselves with no other choice than to accept foreign investment in hazardous industries. Commodity prices have plummeted and debts have risen. In this climate, developing countries have been desperate for foreign investment. For those countries undertaking economic adjustment programs, there has been a liberalization of foreign investment codes encouraged by global economic institutions such as the IMF and World Bank. One of the means of attracting investment in a climate where many developing countries are doing the same has been to relax the enforcement of environmental regulations, which gives investors the impression that their costs will be lower in those countries.

Regardless of the relative weight of the various explanations, the costs incurred by firms with respect to the environment appear to be lower in developing countries than those of the home country. Though environmental cost differences may still seem to be small in absolute terms, they can be extremely important in firms' decisions of where to locate. Global economic competition today is much more fierce than it was in the 1970s, when many of the earlier studies on the issue were undertaken.[23] Firms can

---

[21] Gareth Porter, "Trade Competition and Pollution Standards: 'Race to the Bottom' or 'Stuck at the Bottom'?" *Journal of Environment and Development* 8, no. 2 (1999): 133–51.
[22] Ibid., 136.
[23] Atkinson 1996, 129.

respond to even small differences in costs much more fluidly and rapidly than ever before. There is thus a need for a more thorough investigation into the migration of hazardous industries from richer to poorer countries that takes all of these cost factors into account.

Recent Trends in Toxic Investment

Whatever the degree of influence of environmental push and pull factors on hazardous industry location, it is widely acknowledged that pollution intensity is growing in developing countries.[24] In many instances the hazardous production processes of transnational firms have already relocated to developing countries and have had a detrimental environmental impact. Environmental groups have brought many of these cases to the public's attention. Three trends in the transfer of hazardous industrial processes can be identified. First is the growing amount of foreign direct investment in hazardous industries in developing countries relative to industrialized countries. Second, TNCs often practice double standards. Once they have set up shop in developing countries, they take advantage of lax environmental regulations, whereas similar firms in the home country follow stricter standards.[25] A third trend is the sale of outdated and hazardous plant equipment to local firms in developing countries.[26] While many acknowledge the emergence of these trends, a number of economic studies have argued that TNCs are often cleaner than their local counterparts, even though their performance in developing countries may be less environmentally sound than at home.[27] At the same time, however, it can be shown that TNCs have contributed to hazardous pollution in developing countries and they have not been good role models for local firms. Nor have they been very effective in transferring the most environmentally sound technologies to developing countries.

The share of hazardous and pollution-intensive industries in foreign direct investment in the early 1990s was between 20 and 50 percent in

---

[24] Patrick Low, "International Trade and the Environment: An Overview," in Low 1992, 3.
[25] Castleman 1985; UN TNC and Management Division 1992, 223.
[26] Ian Anderson, "Dangerous Technology Dumped on Third World," *New Scientist* 133 (March 7, 1992): 9.
[27] Norman Bailey, "Foreign Direct Investment and Environmental Protection in the Third World," in *Trade and the Environment: Law, Economics and Policy*, ed. Durwood Zaelke, Paul Orbuch, and Robert Housman (Washington, D.C.: Island Press, 1993), 136; Low 1993, 32; Cheryl Hogue, "Multinationals Help Raise Standards in Developing World, Industry Group Says," *International Environment Reporter* 22, no. 5 (March 3, 1999).

developing countries and 30 and 60 percent in developed countries.[28] But developing countries saw their share rise in the 1980s, while at the same time the rich countries' share (the United Kingdom excepted) declined.[29] And there appears to be widespread evidence that TNCs in the most hazardous sectors of industry that have set up shop in the poorer countries *do* take advantage of lower environmental regulations and of the relatively low level of public protest against polluting activities. For example, according to one UN study, over half of the transnational firms surveyed in the Asia-Pacific region followed standards that were lower than those to which they adhered in developed countries.[30] Such double standards, referred to by some as a sin of evasion, can be applied not only to firms producing similar products in two or more plants in different locations but also to firms with plants in one location that fail to adopt the best available technology for environmental performance.[31] This practice appears to be widespread. One UN study found that less than half of TNCs surveyed had procedures in place for coordinating environmental health and safety policies at both home and in fully controlled foreign affiliates, and only 15 percent of firms with partially controlled joint ventures do so.[32]

In line with these trends, environmental activists have identified a growing number of individual cases of foreign direct investment in the most hazardous industries in developing countries. Earlier cases of the relocation of certain very hazardous industries and the practice of double standards and sin of evasion by TNCs in developing countries have been outlined in detail elsewhere.[33] Below are just a few of the more recent and

---

[28] UN TNC and Management Division 1992, 231.
[29] Ibid.
[30] ESCAP/UNCTC, *Environmental Aspects of Transnational Corporation Activities in Pollution-Intensive Industries in Selected Asian and Pacific Developing Countries* (Bangkok: U.N./ESCAP, 1990), 61.
[31] Birtha Bergsto and Sylvi Endresen, "From North to South: A Locational Shift in Industrial Pollution?" (FIL Working Paper 6), in *Industrial Pollution in the South*, ed. Bergsto et al. (Oslo: FIL, 1995), 19.
[32] UNCTAD Program on Transnational Corporations, *Environmental Management in Transnational Corporations: Report on the Benchmark Corporate Environmental Survey*, Environment Series 4 (New York: United Nations, 1993), 60.
[33] See, for example, Scott Frey, "The Export of Hazardous Industries to the Peripheral Zones of the World System," *Journal of Developing Societies* 14, no. 1 (1998): 66–81; Castleman 1979; Castleman 1985; Rene Ofreneo, "Japan and the Environmental Degradation of the Philippines," in *Asia's Environmental Crisis*, ed. Michael Howard (Boulder: Westview, 1993); Jun Nishikawa, "The Strategy of Japanese Multinationals and Southeast Asia," in *Development and the Environmental Crisis: A Malaysian Case*, ed. Consumer's Association of Penang (Penang: CAP, 1982); Thomas Gladwin, "A Case Study of the Bhopal Tragedy," in Pearson 1987.

ongoing examples in developing countries. Though these cases do not provide statistical evidence for this trend, and for this reason they may not convince economists wedded to the common wisdom, they do indicate that the problem does exist for the most hazardous industries and is prevalent enough to take notice of.

The *maquiladora* firms in Mexico have been cited in recent years as one of the clearest cases of the movement of hazardous industries with the purpose of avoiding strict environmental legislation and enforcement. These U.S.-owned industrial factories located just inside the Mexican border were set up in the 1960s to produce goods for export to the United States. Under a 1983 U.S.-Mexico environmental cooperation agreement, the La Paz Agreement, and under Mexican law, waste generated in these firms must be returned to the country where the raw materials originated, primarily the United States, for disposal.[34] In the 1960s and 1970s most of the *maquiladora* factories were in sectors such as garment assembly and did not produce large amounts of hazardous wastes. In the past decade, however, the composition of *maquiladora* factories has changed dramatically, with the main sectors being chemicals, electronics, and furniture, all of which generate large amounts of toxic waste.[35] By the early 1990s some 87 percent of *maquiladoras* used toxic materials in their production. Moreover, the quantity of investment in these sectors has grown exponentially since the 1970s.[36] Illustrating this qualitative and quantitative shift, investment in Mexican *maquiladora* factories in the chemicals sector, for example, increased by over twenty-fold between 1982 and 1990. This trend followed a tightening of environmental regulations on the industry in the United States.[37] But despite this rapid growth in the generation of hazardous wastes in Mexico, in the early 1990s it was estimated that only about 2 to 3 percent of firms bound by the La Paz Agreement actually returned their wastes to the United States.[38]

---

[34] See Barbara Scramstad, "Transboundary Movement of Hazardous Waste from the United States to Mexico," *Transnational Lawyer* 4 (1991): 253–90.
[35] Edward Williams, "The Maquiladora Industry and Environmental Degradation in the United States–Mexico Borderlands," *St. Mary's Law Journal* 27, no. 4 (1996): 777–79.
[36] Leslie Sklair, *Assembling for Development* (San Diego: University of California, Center for U.S.-Mexican Studies, 1993), 79–80.
[37] David Molina, "A Comment on Whether Maquiladoras Are in Mexico for Low Wages or to Avoid Pollution Abatement Costs," *Journal of Environment and Development* 2, no. 1 (1993): 232.
[38] Sklair 1993, 253–54; Molina 1993, 227–28. Diane Perry et al., "Binational Management of Hazardous Waste: The Maquiladora Industry at the US-Mexico Border," *Environmental Management* 14, no. 4 (1990): 442.

It appears that Mexico was seen as a pollution haven for industries in the United States seeking to escape more stringent environmental regulations at home. Scott Frey notes that according to one survey of U.S. firms with facilities in Mexicali, Mexico, 25 percent of respondents cited lax environmental regulations as a key influence in their decision to locate there.[39] Another survey, conducted by the U.S. Government Accounting Office, indicated that 1 to 3 percent of wood furniture manufacturers relocated their operations from Los Angeles to Mexico between 1988 and 1990 following the adoption of more stringent regulations on air pollution in California. Eighty percent of those firms cited environmental costs in their decision to relocate.[40] This finding confirms Leslie Sklair's observation that "there is little doubt that some U.S. manufacturers have established maquilas in order to escape strict U.S. environmental regulations, including the expensive toxic waste regulations."[41]

The North American Free Trade Agreement (NAFTA), which was negotiated in the early 1990s and came into effect on January 1, 1994, recognized the importance of hazardous waste issues. The agreement specifically mentions the Basel Convention, along with several other environmental agreements with trade provisions, as taking precedence over the NAFTA when there are inconsistencies between them.[42] It also specifically asks the signatories to promise not to use relaxed environmental standards as a means by which to attract foreign investment. How this latter provision is to be enforced, however, is not clear.

In the run-up to the ratification of the NAFTA, both the United States and Mexico had admitted that they did not know the exact the amount of hazardous waste produced by the *maquiladoras* and believed that it was much more than recorded figures indicated.[43] In response to this problem, the United States and Mexico in 1993 set up the U.S.-Mexico Integrated Border Environmental Plan. This plan included the establishment of a computer database of hazardous waste movements between the United States and Mexico called HAZTRAKS. This system has improved the reliability of data, but gaps and inconsistencies still exist. It is widely accepted

---

[39] Frey 1998, 70.
[40] Molina 1993, 227.
[41] Sklair 1993, 254.
[42] Stephen Lerner, "The Maquiladoras and Hazardous Waste: The Effects under NAFTA," *Transnational Lawyer* 6 (1993): 262–64.
[43] John Harbison and Taunya McLarty, "A Move Away from the Moral Arbitrariness of Maquila and NAFTA-Related Toxic Harms," *UCLA Journal of Environmental Law and Policy* 14, no. 1 (1995–96): 6.

that serious mishandling of wastes and illegal dumping in the Mexico-U.S. border region by these firms continues, though more recent figures indicate that around 25–30 percent of *maquila* firms now return toxic waste to the Unites States.[44] It is estimated that these firms produced in the range of 165 tons of hazardous waste per day in 1995, of which approximately 44 tons are unaccounted for.[45] Illegal dumping of toxic waste has been rampant, and it is estimated that there are hundreds of potentially toxic waste dumps along the border.[46]

While there is indeed a serious problem in the *maquiladora* industries in Mexico, the Basel Ban Amendment, adopted in 1995, will not apply to waste movements between the United States and Mexico because Mexico joined the OECD in 1994. Mexico is thus an Annex VII country and is allowed to trade hazardous wastes freely with other Annex VII countries, subject to the other provisions of the convention. Many saw the move to allow Mexico to join the OECD as primarily politically motivated to calm U.S. concerns at the time that Mexico, as a developing country, might unfairly take jobs from the United States once NAFTA came into force. But Mexico is still considered by many to be a developing country, and the conditions along the border region only serve to confirm this assessment.

The movement of hazardous industries to developing countries, which is evident in the *maquiladoras* in Mexico, is also happening on a global scale. Environmental groups have highlighted these cases in a bid to raise awareness of this practice. TNCs in the chemicals industry in particular are in the process of relocating their production overseas to Asia, the Pacific Rim, and Latin America. This has occurred just as the industry faced a drop in demand in the West, while demand in the newly industrializing countries has rapidly increased.[47] Rather than export their chemical products to these newly industrializing regions, these firms have chosen to set up new facilities closer to the market. But this decision was made not just to save on labor and transport costs. Environmental reasons have also played a role. Increasingly stringent environmental regulations in Europe,

---

[44] Cyrus Reed, "Hazardous Waste Management on the Border: Problems with Practices and Oversight Continue," *Borderlines* 6, no. 5 (July 1998). See also U.S. EPA HAZTRAKS at http://www.epa.gov/earth116/6en/h/haztraks/haztraks.htm.

[45] Reed 1998. See also "Free Trade, Hazardous Waste," *Borderlines* 5, no. 6 (June 1997). http://www.zianet.com/irc1/borderline/1997/bl136/bl136haz.html.

[46] Enrique Medina, "Overview of Transboundary Pollution Issues along the Mexico-US Border," in *Environmental Toxicology and Risk Assessment: Fourth Volume*, ed. Thomas La Point, Fred Price, and Edward Little, ASTM STP 1262 (West Conshohocken, Penn.: American Society for Testing and Materials, 1996), 9.

[47] Paul Abrahams, "The Dye Is Cast by Growth and Costs," *Financial Times*, May 31, 1994.

for example, have been cited as a main contributor to the movement of production facilities to Asia.[48] The chief executive of Bayer in 1994, Manfred Schneider, for example, has stated regarding relocation to Asia: "The main disadvantages we have to face are higher labor costs and expensive social security systems, coupled with widespread regulation of environmental affairs by the state. We have to overcome these handicaps. If we do not, many areas of our business [in Europe] will become uncompetitive and therefore in danger of being squeezed out of the market."[49]

Many of these global chemicals firms operating in developing countries produce highly toxic substances and wastes, particularly those containing chlorine. As the industrialized countries are implementing a phaseout of chlorofluorocarbons (CFCs) to comply with the Montreal Protocol to Protect the Ozone Layer, economic decline has beset the chlorine industry.[50] This has contributed to the shift in production to poorer countries, where demand for chlorine is still high. For example, chemical companies are encouraging the production of polyvinyl chloride (PVC), of which chlorine is a key ingredient, as an environmentally friendly alternative to wood in developing countries. But products containing chlorine and which generate chlorine wastes are highly dangerous, both when disposed of in landfills and when incinerated. The latter process is particularly toxic because it releases dioxins and furans, some of the most hazardous chemicals known to humans. For this reason, the U.S. government has banned the land disposal of chlorine wastes and has placed a moratorium on the incineration of hazardous wastes. This tightened regulation on the chlorine industry appears to have contributed to the exodus of these firms to developing countries.

A striking example of the pollution resulting from the chlorine industry's operations in less industrialized countries is the case of a subsidiary of the U.S.-based Pennwalt Corporation located in Managua, Nicaragua. First set up in 1968, this plant produced chloralkalai using a mercury cell process. By the early 1990s, the firm had dumped over ninety tons of mercury in Lake Managua. Mercury poisoning afflicted a number of workers at the plant.[51] The health of residents near the plant was also compromised

---

[48] Andrew Wood, "Asia-Pacific: Rising Star on the Chemical Stage," *Chemical Week*, 156 no. 6 (February 15, 1995), 36.
[49] Quoted in Abrahams 1994.
[50] Kenny Bruno and Jed Greer, "Chlorine Chemistry Expansion: The Environmental Mistake of the 21st Century," *Toxic Trade Update* 6, no. 2 (1993): 1–3; Joshua Karliner, "The Environmental Industry," *Ecologist* 24, no. 2 (1994): 61.
[51] "Niagara to Nicaragua," *Waste Trade Update* 5, no. 1 (1992): 4–6; "Update on Elpesa," *Toxic Trade Update* 5, no. 2 (1992): 25.

by the regular release of chlorine gas. The mercury cell process for producing chlorine at the Managua plant continued into the 1990s, though a similar chloralkalai plant owned by Pennwalt in Niagara Falls, New York, switched to a less polluting process in 1987. Numerous local groups waged a campaign to close the Managua plant. It was finally ordered to shut down in 1992, not for environmental reasons but because the firm was unable to repay its debts to a regional development bank.[52]

Subsidiaries of multinationals in the chemicals industry have taken advantage of lax environmental regulations and poor enforcement of those regulations in the developing world. In some cases, though, these firms have been ordered to stop production in response to environmental and health complaints. In one case, a subsidiary of Rhone-Poulenc, a French chemical firm, was accused by the government of Brazil of dumping toxic waste in that country between 1966 and 1984. It was also accused of exposing workers to dangerous levels of toxic residues at its solvents plant, which was shut down in 1993. More than 150 workers at this firm had levels of hexachlorobenzine (HBC) which were far in excess of that considered to be dangerous. In an attempt to clean up the problem, the company installed an incinerator to burn the toxic waste.[53]

In Papan, Malaysia, a "rare earth" plant was set up in 1982 by a firm owned in part by the Japanese multinational Mitsubishi. This plant extracted highly toxic compounds for the production of electronic components such as color television screens. Radioactive wastes from the plant were indiscriminately dumped in plastic bags behind the plant.[54] The wastes severely contaminated the area. When tested in 1984, it was found to have radiation levels that were well above internationally acceptable levels.[55] Since the firm opened, the rate of illness in nearby villages has increased dramatically.[56] Members of the community took legal action against Mitsubishi.[57] In 1992 the plant was ordered by a Malaysian court to stop production, but the firm appealed the decision.

---

[52] "Elpesa Closed by Development Bank," *Toxic Trade Update* 6, no. 1 (1993): 32–33.
[53] Patrick McCurry, "Brazil Warns Rhone-Poulenc on Toxic Waste," *Financial Times*, November 8, 1994.
[54] Karliner 1994, 61.
[55] Castleman 1987, 164.
[56] "Malaysian Villages Welcome Court Ruling to Close Mitsubishi/ARE Facility," *Toxic Trade Update* 5, no. 2 (1992): 24.
[57] Mike Ichihara and Andrew Harding, "Human Rights, the Environment and Radioactive Waste: A Study of the Asian Rare Earth Case in Malaysia," *Review of European Community and International Environmental Law* 4, no. 1 (1995): 1–14.

In the early 1990s, Bayer faced protests at its subsidiary chromium plant in Durban, South Africa. Highly toxic wastes from the plant had been plowed into the earth in a dump site near residential areas and had contaminated a nearby canal. Workers at the plant were not warned of the dangers of working with chromium. Increased health problems and several deaths from chrome exposure resulted from the poor conditions at the plant. As a result of these problems, the plant was forced to close in 1992.[58]

Some foreign investors in the developing world have actively sought to avoid environmental liability. DuPont, for example, in setting up a nylon factory in Goa, India, during the late 1980s and early 1990s, negotiated a limited liability clause with its local partner that exempts DuPont from liability in the case of a chemical accident or pollution. Environmental activists have argued that this is a clear attempt to try to avoid the legal hassles that Union Carbide faced in the wake of the explosion at its pesticides plant in Bhopal, India, in 1983. The Bhopal incident, itself a striking case of double standards, gave Union Carbide bad press for years as it attempted to extract itself from any liability or compensation to victims. The construction of the DuPont nylon factory at Goa in the 1990s prompted protests from local residents who demanded that the plant be shut down.[59]

On top of the continued growth in foreign direct investment in hazardous industries and the related problem of double standards is the growing incidence of the sale of entire "used" manufacturing plants from industrialized countries to poorer countries. Many of these sales involve the transfer of outdated technologies that are no longer wanted because they are either judged to be inefficient or involve hazardous processes that rich industrialized countries are seeking to replace to meet rising environmental standards on industry at home.[60] The movement of these plants to contexts that they were not originally planned for can lead to problems, especially when they are sold to developing countries that have few facilities capable of treating the wastes these plants produce. Not only does this practice result in the transfer of environmentally inferior technology, but, ironically, the subsidiaries of the big multinationals in those countries come out looking relatively clean by comparison.

---

[58] "Bayer Poisons South Africa," *Toxic Trade Update* 5, no. 2 (1992): 23.
[59] Gary Cohen and Satinath Sarangi, "DuPont: Spinning Its Wheels in India," *Multinational Monitor* 16, no. 3 (1995): 23.
[60] Andrew Taylor, "Third World Looks to First World Cast-Offs," *Financial Times*, January 16, 1996, 5.

Most of the outdated plants and other equipment that are sold to developing countries are in the manufacturing and chemical industries. According to one of the partners of a international plant and property consulting firm that handles such sales, much of it has made its way to Asia, particularly China, the Indian subcontinent, and Southeast Asia.[61] There is a high demand in these countries for this equipment, as it is seen as an affordable way to speed up industrialization. Examples of outdated and environmentally questionable technology transfer to developing countries include the sales of secondhand plants involving chlorine and mercury. In 1994 an Indian firm, United Phosphorus, purchased a secondhand chlorine production plant from the Norwegian firm Norske Skog after the Norwegian government adopted a policy to reduce organochlorine production. This sale caused a stir among protesting environmental groups, but the Indian firm has carried through on the deal.[62] The plant plans to produce highly toxic agrochemicals in India.

Also in 1994, a Danish firm, Dansk Sojakagen Industries, attempted to sell a contaminated secondhand mercury-based chloralkalai plant to Ravi Alkalis, a Pakistani alkali firm. If it were disposed of in Denmark, the plant would technically be deemed as hazardous waste. But because it was sold as equipment, the sale was not technically illegal under the Basel Convention. After the deal was publicized by Greenpeace and other environmental groups, the Pakistani firm agreed to discard the part of the plant equipment involved in hazardous production processes.[63] The governments of both Denmark and Pakistan were clearly embarrassed by the incident. In the past decade, other countries, including Indonesia, have also been the recipients of secondhand chloralkalai technology, which is much less expensive to purchase than are newer, cleaner technologies.[64] Some seventy-four mercury cell process plants are due to be closed down by the year 2010. Environmental groups have called for industrialized countries to track mercury cell as well as other toxic technologies to prevent their shipment to developing countries.[65]

---

[61] Ibid.
[62] "Indian Firm Sues Journalists over Chlorine Plant," *International Toxics Investigator* 8, no. 1 (1996): 8.
[63] Kenny Bruno and Janus Hillgaard, "Denmark Dumping Chlorine Technology in Pakistan," *Toxic Trade Update* 7, no. 1 (1994): 42–43; Kenny Bruno, "Rejecting Toxic Technology," *Multinational Monitor*, 16 no. 1/2 (1995), http://www.essential.org/monitor/hyper/mm0195.html#environ.
[64] Anderson 1992, 9.
[65] Bruno 1995.

Government-funded export credit agencies have been under fire in recent years from environmental NGOs over the funding and provision of insurance for environmentally dubious investments in developing countries, including chemical facilities and other hazardous manufacturing plants.[66] An official with the U.S. Export-Import Bank recently stated that "we are funding all sorts of projects within the chemical group, large and small."[67] Data on funding for particular cases can be found on the websites of some of the more transparent export credit agencies.[68] These include, for example, funding for a battery manufacturing facility in Angola, and insurance for the sale of used chemical processing equipment to a firm in Mexico. Most export credit agencies have been very secretive in their operations, however, making it difficult to know the full extent of funding for such investments overseas.[69]

Do these various cases constitute sound proof that hazardous firms are relocating to developing countries for environmental reasons? It is difficult to prove definitively that they do without detailed information about firms' total environmental costs and true motivations for location decisions, data that are not readily available. But these various cases are consistent with other studies, cited above, that have found a trend of relocation of the most hazardous industries to developing countries and the practice of double standards in the past. Moreover, environmental costs for firms appear to be rising in OECD as opposed to non-OECD countries. In today's increasingly global economy, such price differentials may indeed constitute a strong motivation for industry relocation, and this may increase in the future.

Pollution Intensity of Local Industry versus TNCs

Some have argued that the developing world's growing share of pollution-intensive industry has been largely the result of increased domestic

---

[66] See ECA Watch website: www.eca-watch.org/introduction.html.
[67] Transcript of Ex-Im Bank U.S.A. Chairman James A. Harmon Question and Answer Session Following Environment Speech—Mumbai, February 24, 2000, at http://www.exim.gov/press/feb2500.html.
[68] These include the Overseas Private Investment Corporation (OPIC) at www.opic.gov and the U.S. Export-Import Bank at www.exim.gov.
[69] "A Race to the Bottom: Creating Risk, Generating Debt, and Guaranteeing Environmental Destruction," A Report by Berne Declaration, Switzerland; Bioforum, Indonesia; Center for International Environmental Law, U.S.; Environmental Defense, U.S.; Eurodad, Belgium; Friends of the Earth, U.S.; Pacific Environment and Resources Center, U.S.; Urgewald, Germany, at http://www.environmentaldefense.org/programs/International/ECR/race.html.

demand in these countries for pollution-intensive products, and that foreign investment has played only a minor role.[70] It is also widely argued that TNCs, even in polluting industries, are likely to use more environmentally sound technologies than local firms because they have access to better technology and anticipate stricter environmental laws in future.[71] Indeed, some stress that lax environmental regulations are actually a deterrent to foreign investment.[72] These are important points, which require further examination. Data on the generation of hazardous wastes by locally owned firms, particularly small and medium-sized enterprises in developing countries, are extremely sparse. Below I discuss this question with particular reference to Southeast Asia, an area on which most information on this issue can be found.

The main domestic industries responsible for toxic emissions in Southeast Asia include leather tanning, pesticides, chemicals, metal plating, electronic industries, and textile dyeing. The prevalence of these industries varies by country. Most of the wastes generated by locally owned firms in these industries are indiscriminately dumped into rivers, mixed with domestic solid waste and disposed of in landfills, or stored on site. The lack of sufficient facilities to treat hazardous waste has only compounded this problem.

Most locally owned firms in Southeast Asia are small and medium-scale enterprises (SMEs). Locally owned and operated SMEs are likely responsible for a portion of hazardous waste generation and improper disposal in Southeast Asia, particularly those in the leather tanning, textile dyeing, and metal plating industries. But much of the waste generated by small and medium-sized local firms is tied to global economic relationships (for example, export industries such as computer chips, electroplating, and textile dyeing) that are increasingly becoming integrated into the global economy. Though domestic demand for such products may be rising at the same time, these industries are also clearly geared toward export markets.

The argument is often put forward that these small and medium-scale domestic firms are more polluting than larger and multinational-affiliated firms because they lack funds to install cleaner production technologies and to dispose properly of the hazardous wastes they generate. Locally owned firms may also be more polluting because incentives to be more clean are lower where land ownership and rights are not clearly delineated,

---

[70] Mani and Wheeler 1998.
[71] Low 1993, 32; Birdsall and Wheeler 1992; Ferrantino 1997, 55–56.
[72] Repetto 1994, 23.

as is the case in much of Southeast Asia.[73] In addition, the small size of SMEs and the fact that they are often dispersed may make it easier for them to evade government regulations.

There are a variety of reasons for pollution problems in locally owned firms, large and small, in Southeast Asia. Many locally owned firms do not have any policies on pollution control, in contrast to most TNCs and their affiliates currently operating in the region, which do at least have policies related to pollution control. For example, one UN study indicated that only half of the locally owned firms surveyed in the Asian semiconductor industry had any environmental policy at all. The report stated that where such policies existed, they were usually based on local laws, themselves often weak and not enforced.[74] Compliance with local laws among locally owned firms is in many cases extremely low. For example, in the leather tanning industry in Indonesia, compliance with pollution discharge regulations in the early 1990s was only 2 percent.[75] This poor performance reflects the fact that locally owned firms often lack monitoring equipment and have poor training with respect to environmental concerns. Health risks to workers in locally owned firms are also high because of the inadequate availability of safety equipment and a lack of adequate emergency procedures.[76]

These problems are compounded by a lack of industrial zoning policies in most Southeast Asian countries. Industrial firms of all sizes, especially smaller firms, are often located near residential areas. Poor urban communities are often most affected, with large shantytowns located near industrial areas. The lack of separation between residential and industrial areas increases the exposure of urban populations to toxins from hazardous wastes and industrial effluents, particularly along rivers, where toxic waste is often discharged by local manufacturing firms.

One of the largest problems faced by local firms with respect to environmental policies and procedures is their lack of financial resources with which to develop or operate elaborate pollution control policies or technologies.[77] SMEs in particular often do not have the funds to hire environmental

---

[73] Lawrence Kent, *The Relationship between Small Enterprises and Environmental Degradation in the Developing World, with Special Emphasis on Asia* (Washington, D.C.: Development Alternatives, 1991), 10.
[74] ESCAP/UNCTC 1990, 69–70.
[75] Carter Brandon and Ramesh Ramankutty, *Toward an Environmental Strategy for Asia*, World Bank Discussion Paper 224 (Washington, D.C.: World Bank, 1993), 69.
[76] ESCAP/UNCTC 1990, 71–72.
[77] David O'Connor, *Managing the Environment with Rapid Industrialisation: Lessons from the East Asian Experience* (Paris: OECD, 1994), 167–72.

consultants to help them improve their environmental performance, and they find it especially difficult to take advantage of economies of scale relating to the adoption of environmentally sound technologies.

While these problems are indeed significant in Southeast Asia, it is also clear that large firms, many of which are TNCs, also play an important role in the generation of toxic wastes in the region. In assessing the role of SMEs in the toxic waste crisis, it is important to determine their share of manufacturing output and the amount of waste generated per unit of output as compared to larger firms.[78] In some parts of Asia, such as India and China, SMEs make up 60–70 percent of manufacturing output, and thus their contribution to the toxic waste crisis is significant. But in other Asian countries, such as those in Southeast Asia, the share of SMEs in total output is much lower. For example, small and medium-sized firms make up 18 percent of output in Indonesia and about 13 percent in the Philippines.[79] An illustration is the leather tanning industry in Indonesia, which consists of some 540 enterprises. The vast majority, 470 firms, are small enterprises that make up only 12 percent of the total output. The World Bank notes that "although [SMEs] are not the major polluters in most subsectors, they often pollute more per unit of output than large firms operating in the same sector."[80]

So though SMEs may pollute more per unit of manufacturing output than large firms, they do so in relatively small amounts because their output is much smaller compared to large firms, many of which are geared toward export markets or are TNC affiliates. SMEs, then, apparently are not necessarily the most significant source of toxic waste in the region, but they do contribute to the problem. Because the share of production made up by domestic industries varies by country, it is difficult to generalize that TNCs are not the main source of toxic pollution in the region or for the developing world as a whole. Both local production and TNC production in toxic industries in the developing world pose serious problems, and both must be taken into account.

Conclusion

I have argued in this chapter that there appears to be a distinct and ongoing pattern of investment by TNCs in hazardous industries in developing

---

[78] Kent 1991, 13–22.
[79] Brandon and Ramankutty 1993, 74.
[80] Ibid., 72.

countries. This trend represents yet another avenue for hazard transfer. There are reasons to expect that this practice will likely continue given changes in the current global economic and regulatory context. These changes, including the new global rules on the hazardous waste trade which are part of the Basel Convention, and more fierce global economic competition among firms, have led to rising environmental costs faced by industrialized country firms in hazardous industries. At the same time, there has been an increased sensitivity to cost differentials between rich and poor countries. The failure of many analysts today to recognize these changed global economic and regulatory circumstances has led to an underestimation of the cost differentials between rich and poor countries and an underestimation of their significance to firms' decisions on where to invest. There is a need for further study on the issue of hazardous industry location that takes into account the full environmental costs faced by TNCs and their significance to investment decisions.

Environmental groups have highlighted many cases of hazardous industry migration to developing countries in an attempt to bring the issue to the public's attention. Efforts to elicit change in firms' behavior on this issue, however, have been much more difficult for these groups than has been the case with the waste trade. In the case of the latter, environmental groups could appeal to a global treaty as a standard by which firms and countries should abide. With respect to investment in toxic industries, there is much less international agreement that the practice should be governed by international rules. This problem is no doubt linked to the sparseness of data and uncertainties regarding motives for relocation.

The growing toxicity of industry in developing countries is no doubt also linked in some ways to its general pattern and phase of industrialization. Small and medium sized enterprises, which make up the bulk of domestic industry, may indeed pollute more per unit of output than larger firms. But their share of total output is in many cases small compared to that of the larger firms that are TNC affiliates or geared to the export market. Industrial relocation from rich to poor countries only compounds the problem. Whichever is the major cause, the end result is the same. Toxic waste generation is a serious and growing problem in developing countries at the same time that it is diminishing as a problem in industrialized countries. Moreover, despite campaigning by environmental groups on the need to promote clean production, the actions taken by private industry and governments to tackle the situation have been far from adequate. This issue is explored more fully in the next chapter.

6
Market-Based and Voluntary Initiatives: Promoting Clean Production?

There is wide acknowledgment, even in the international business community, that in the past TNCs in hazardous industries have at times engaged in environmentally unsound activities in less developed host countries. But by the early 1990s industry advocates began to argue that this poor environmental behavior was in the past, and that the performance of TNCs on that front was improving rapidly or at least set to do so.[1] Moreover, many in the business community have argued that TNCs are in an ideal position to transfer clean, rather than hazardous,

---

[1] Stephan Schmidheiny and WBCSD, *Changing Course* (Cambridge, Mass.: MIT Press, 1992); Ann Rappaport, *Development and Transfer of Pollution Prevention Technology* (Westport, Conn.: Quorum Books, 1993).

production technologies to developing countries because they can afford to develop such technologies, and are likely to install them in their affiliates abroad before local firms do.[2] At the 1992 United Nations Conference on Environment and Development, governments called on TNCs to take an active role in promoting clean technology development and transfer. TNCs have indeed taken the lead on these issues in recent years. Rather than waiting for further government regulation, they have taken matters into their own hands. They have made large investments in the global "green" technology market and have established environmental self-regulations.

In this chapter I evaluate the progress on this front over the last decade. I argue that the focus on market-based and voluntary initiatives has resulted in weak mechanisms and little truly clean technology transfer to developing countries because certain features of the global economy, particularly financial globalization, dampen incentives for transfer of clean technology. At the same time, putting these efforts in the hands of industry has contributed to the "privatization of environmental governance" whereby environmental NGOs and developing countries are largely left out of the decision-making process. Because they have extensive lobbying efforts and influence in industrialized countries, private business actors have been able to shape global rules that ostensibly promote clean production in ways that leave the door open for a continuation of hazard transfer.

Clean Production and Transnational Corporations

"Clean production" is generally agreed to refer to methods of production that prevent pollution. Several terms are used in the literature on pollution prevention: clean production, cleaner production, environmentally sound technologies, environmentally sound management, and ecoefficiency. Though there are some slight differences in the definitions of these terms (for example, clean production implies in its most extreme form zero waste, whereas cleaner production implies waste reduction), they largely refer to similar principles.[3] These principles include production technologies that prevent the generation of pollution, partially or entirely.

---

[2] Schmidheiny and WBCSD 1992; William F. Wescott II, "Environmental Technology Co-operation: A Quid Pro Quo for Transnational Corporations and Developing Countries," *Columbia Journal of World Business* 27 (Fall–Winter 1992): 145–53.

[3] On some of the differences in definitions of these terms, see OECD, "Effective Technology Transfer, Co-operation and Capacity Building for Sustainable Development: Common Reference Paper no. 75" (Paris: OECD, 1994), 14–15.

This means technologies that make less intensive use of natural resources, are energy efficient, eliminate the use of toxic raw materials, and eliminate or at least reduce the quantities and toxicity of wastes at all stages of the product's life cycle.[4] This is in contrast to "end of the pipe," "pollution control," or "cleanup" technologies, which focus on the installation of equipment designed to treat wastes and pollution after they have already been generated.

The adoption of clean production technologies by global firms in their operations around the world could go a long way toward reducing hazardous waste exports and incidences of hazardous industry migration and double standards. The United Nations Environment Programme launched a "Cleaner Production Programme" in 1990 to promote cleaner production among firms and governments and to facilitate the transfer of cleaner production technologies globally.[5] The importance of clean production was also recognized at the Earth Summit, and the issue of clean technology transfer was prominent in Agenda 21. This document set goals to strive for on the issue of industrial hazards, particularly in its chapters on hazardous waste, industry, and technology transfer.[6] It asked all countries to reduce the generation of hazardous wastes as part of an integrated cleaner production approach. Governments were asked to regulate industry in such a manner as to prevent industrial hazards such as toxic waste from causing environmental harm.[7] Business and industry were asked explicitly to cooperate with governments by transferring clean production technologies.[8] These new, environmentally friendly industrial technologies were seen as vital for sustainable development, and TNCs were seen as the key channel for their transfer, particularly when they are subject to patent protection.[9] Agenda 21 specifically asks TNCs to "adopt standards of operation with reference to hazardous waste generation and disposal that are equivalent to or no less stringent than standards in the country of origin."[10] The document also sees an important role for voluntary environmental measures on

---

[4] UNEP, "Cleaner Production," http://www.uneptie.org/Cp2/; OECD, *Promoting Cleaner Production in Developing Countries* (Paris: OECD, 1995), 13; Kenneth Green and Alan Irwin, "Clean Technologies," in *The Greening of Industry: Resource Guide and Bibliography*, ed. Peter Groenenegen et al. (Washington, D.C.: Island Press, 1996), 169–94.

[5] UNEP, "Cleaner Production," *UNEP Industry and Environment* 17, no. 4 (October–December 1994): 4.

[6] UN, *Agenda 21* (New York: United Nations, 1992). See especially chaps. 20, 30, and 34.

[7] Ibid., 198–99.

[8] Ibid., 237.

[9] Ibid., 252–53.

[10] Ibid., chap. 20 (especially paragraphs 11, 13, and 29) and chap. 34.

the part of industry as a way to meet these goals of cleaner production and waste minimization. It strongly encourages voluntary efforts by private business and industry associations as a way to meet goals such as hazardous waste reduction and clean technology transfer.[11]

The adoption of clean production methods requires a major shift in the way firms produce their products. The reasons why clean technologies had not yet been adopted or widely transferred to the developing world by the time of the Earth Summit have been widely acknowledged in the business and environment literature. They include, first, high initial cost outlays for firms. These must be paid up front, even though the financial benefits appear in the longer term. In developing countries, securing such financing is difficult, and investors often must rely on donor governments or aid agencies to supply credit.[12] Though longer-term financial benefits of cleaner production have been shown by its proponents, many firms have been too concerned with short-run profits to make the initial investment in clean technologies. Also, where there are regulatory differences between countries, there is little incentive for firms to revamp the production processes in countries that do not require them. There have also been problems of vested interests in old technologies and concerns that changes in the production process will affect the quality of the end product.[13] In addition, lack of knowledge of clean production methods combined with the difficulties of measuring their impact have hampered their adoption.

Although these problems have hindered firms' adoption and transfer of clean technologies in their operations around the world, many in the business community argue that the benefits of adopting clean production now outweigh the costs associated with these problems. Clean production is thus seen to be the wave of the future. Increasing pressure on firms is coming from many quarters. Governments, NGOs, and investors, including banking institutions and individuals, are now demanding that firms clean up their acts.[14] As a result, the costs of "business as usual" are rising for firms, and the benefits of adopting clean production look more attractive.

---

[11] Ibid., chap. 30 (especially paragraph 8).
[12] Saleh Hafez, "Financing Cleaner Production in Developing Countries," *UNEP Industry and Environment* 17, no. 4 (October–December 1994): 75–76.
[13] See, for example, Rappaport 1993, 24–25.
[14] Zorraquin Rappaport and Stephan Schmidheiny, *Financing Change: The Financial Community, Eco-Efficiency and Sustainable Development* (Cambridge, Mass.: MIT Press, 1996); John Ganzi, Frances Seymour, and Sandy Buffet with Navroz Dubash, *Leverage for the Environment: A Guide to the Private Financial Services Industry* (Washington, D.C.: World Resources Institute, 1998).

For example, it is argued that for firms that adopt clean production methods, liability costs, the costs of acquiring new capital, and future waste management costs are all lower. In addition, higher returns are expected from increased market share for cleaner products and from an enhanced image for firms that adopt clean production methods.[15] As these changes became evident to industry around the time of UNCED, it pledged that it would take its leadership role seriously in promoting the adoption and transfer of clean production technologies. Industry and governments alike believed that market-based measures and voluntary initiatives were the most promising ways to pursue these goals.[16]

"Green" Investment and the "Environment Industry": Cleaning Up

Since UNCED there has been a growing interest in "green" investment by TNCs, including the greening of their investments in developing countries. This is seen by many as a case in which the market can provide an environmentally sound solution to growing environmental problems around the world. In the early 1990s there were calls for government support for research, development and transfer of environmental technologies through private channels.[17] Industrialized country governments indeed began to promote green technology investment and exports through their aid programs throughout the decade. There was a boom in the 1990s in the environmental goods and services industry, which includes both cleaner production technologies and end of pipe pollution control technologies.[18] This increased attention to and funding for the environment industry was to have a double benefit of improving the environment and giving a boost to donors' own firms in the global environmental industry sector.[19] This focus on private industry as a key channel for environmental

---

[15] Livio DeSimone and Frank Popoff with WBCSD, *Ecoefficiency: The Business Link to Sustainable Development* (Cambridge, Mass.: MIT Press, 1997).
[16] OECD 1995, 16.
[17] See, for example, George Heaton, Robert Repetto, and Rodney Sobin, *Backs to the Future: US Government Policy Toward Environmentally Critical Technology* (Washington, D.C.: World Resources Institute, 1992); Amitav Rath and Brent Herbert-Copley, *Green Technologies for Development: Transfer, Trade and Cooperation* (Ottawa: IDRC, 1993); OECD, "Effective Technology Transfer, Co-operation and Capacity Building for Sustainable Development: Common Reference Paper no. 75" (Paris: OECD, 1994).
[18] See, for example, OECD, *The Global Environmental Goods and Services Industry* (Paris: OECD, 1996).
[19] OECD, *The Environment Industry: The Washington Meeting* (Paris: OECD, 1996).

technology transfer made sense, as most technologies, including environmental ones, are transferred via commercial channels.[20] This seemed to be an excellent method for diffusing green technologies on a global scale, and it was widely assumed that it would also benefit developing countries.

The reality of this green investment strayed from the original goals outlined in Agenda 21. For the most part, the environmental technologies that are being transferred to developing countries are not so much clean production technologies as they are cleanup technologies. The reason appears to be that the latter, particularly in the case of hazardous industry, are more profitable than the former, yet they still enhance firms' green image. The growing generation of hazardous waste in the developing world has provided a market for the increasingly global cleanup industries, which are now investing under the label of "environmental technologies." The promotion of this type of environmental investment by industrialized countries, through aid programs, has also tended to focus on end of the pipe technologies rather than on pollution prevention technologies. This trend has been acknowledged by the OECD.[21]

Much of the cleanup investment in developing countries is being undertaken by subsidiaries of some of the biggest TNCs involved in hazardous industries. For example, global chemical giants Dow, DuPont, and Monsanto have each set up their own international "environmental" firms which specialize in hazardous waste treatment. They argue that branching out into this line of business is a "natural fit."[22] The global cleanup industry, for which the United States is the largest market, grew 20 percent per year in the 1980s and was valued at U.S.$250 billion—U.S.$300 billion annually in the early 1990s.[23] This rapid growth was in response to pressure on firms primarily in industrialized countries to clean up sites that they had contaminated through improper disposal of wastes and to ensure that hazardous waste streams were dealt with legally through licensed toxic waste management firms. This market grew in North America in the 1980s, but began to stagnate in the mid-1990s.[24] This may be accounted for by the installation of cleaner technologies in the face of rising liability

---

[20] OECD 1995, 69.
[21] OECD 1995, 71; OECD, *Cleaner Production and Waste Minimization in OECD and Dynamic Non-Member Economies* (Paris: OECD, 1997), 30–31.
[22] Allison Lucas, "Chemical Companies Play the Environmental Market," *Chemical Week* 156, no. 2 (January 18, 1995): 29–34.
[23] Lucas 1995, 29; Adam Schwarz, "Everybody's Business: Asia Has Plenty of Waste, but Profiting from It Is Tough," *Far Eastern Economic Review*, (November 24, 1994): 136.
[24] Elisabeth Kirschner, "A New World for Environmental Services," *Chemical Week* 154, no. 2 (January 19, 1994): 28–31; Lucas 1995, 29.

costs in the early 1990s. But a more troubling explanation may be that in countries such as Canada and the United States, where government regulations on the environment were scaled back in the mid-1990s, demand for environmental services declined. This did not deter cleanup firms, which then made determined efforts to go global.[25] Investment in the rapidly industrializing countries was seen as a way to boost their drooping profits.

Joshua Karliner has argued that this movement of the cleanup industry is part of a three-stage process. First, toxic industry is exported from rich to poor countries. This is followed by the export of environmental regulations, with industrialized countries offering assistance to developing countries seeking to improve their environmental laws. Finally, there is an export of environmental cleanup technologies that enable waste-producing firms to meet those regulations.[26] When the cleanup firms are subsidiaries of the same corporations that generate large amounts of hazardous waste in developing countries, the double benefits are obvious.

This continued focus on cleanup rather than clean technologies into the 1990s appears to be linked to immediate financial returns. Short-run profits are ultimately higher and relatively more secure in cleanup operations than they are in the development of truly clean technologies because the former are seen to carry less short-term risk. The installation of truly clean technologies that actually eliminate hazardous waste generation involves large initial capital outlays, and any savings that may exist will likely be realized only over the very long run.[27] As finance becomes increasingly global, shareholders and financial institutions demand short-term payoffs, making clean technology investments less attractive to firms. It is not clear that firms will take the risk to transfer clean production technologies, which require much longer time frames than they currently care to undertake, without financial backing from state aid agencies or multilateral institutions. Cleanup investments are risky enough and already rely on such forms of aid.

The rapidly industrializing countries in the Asia-Pacific region are increasingly seen since the 1990s as the most promising markets for environmental cleanup firms that were going global. There is both plenty of

---

[25] Kate O'Neill, "Transnational Trash: The Globalization of the Waste Disposal Industry," paper presented at the International Studies Association Annual Meeting, Toronto, March 18–22, 1997.
[26] Joshua Karliner, "The Environmental Industry," *Ecologist* 24, no. 2 (1994): 60–61. See also Karliner, *The Corporate Planet* (San Francisco: Sierra Club, 1997).
[27] Bernard Fleet and Jeremy Warford, "Managing Hazardous Industrial Wastes in Developing Countries," *Ecodecision* 14 (Fall 1994): 42–43.

hazardous waste to be cleaned up and economic assistance for investors. The sustained economic and industrial activity in the region has generated of a great deal of hazardous waste. Some countries in Southeast Asia, for example, may produce as much or more hazardous waste per kilometer as in the United States.[28] This growth in hazardous waste generation has in turn led some Asian countries to tighten up regulations on waste disposal. This has been good news for the global environment industry, as it has meant a growing market for cleaning up after hazardous industries. As Edwin Falkman, the London-based CEO for the U.S. hazardous waste management firm WMX has remarked, "Asia is one of our major growth areas for the future. Increases in wealth and population across Asia are creating some very significant opportunities."[29]

Until the early to mid-1990s, none of these countries had proper hazardous waste disposal facilities. Yet many firms, both TNCs and local industry, were certainly producing hazardous waste and disposing of it in environmentally unsound ways. Because national environmental regulations in the region have not required hazardous waste to be disposed of separately from other solid wastes, much of it has been dumped into rivers or in municipal garbage dumps. This lack of regulation has led to a growing number of environmental incidents such as the 1995 discovery of forty-one drums of toxic potassium cyanide dumped by a small chemical trading firm on Pangkor Island in Malaysia.[30] An official of the Malaysian International Development Authority remarked that this incident was just the tip of the iceberg. Similar cases have occurred across the developing world. These sites must now be remediated. Hong Kong and Taiwan have stepped up regulation on hazardous waste disposal. Since global firms seem to be the only ones with the technology to equip these disposal operations, they are placing themselves to make the most of this opportunity.

U.S. cleanup firms appear to be the most aggressive in seeking out markets for their services in developing countries, though Japanese and European firms are also staking out their portion of the market.[31] The U.S. firm WMX, for example, expected Asia to account for as much business as Europe by 2000. In recent years WMX has set up a chemical waste

---

[28] Jonathan Lindsay, "Overlaps and Tradeoffs: Coordinating Policies for Sustainable Development in Asia and the Pacific," *Journal of Developing Areas* 28 (October 1993): 28.
[29] Quoted in Schwarz 1994, 136.
[30] S. Jayasankaran, "Waste Not, Want Not: Malaysia Needs a Waste-Treatment Facility Fast," *Far Eastern Economic Review* (April 13, 1995): 61.
[31] Kirschner 1994, 28–31; George Vander Velde and Terre Belt, "Hazardous Waste: Global Roundup," *Engineering News Record*, no. 233 (October 17, 1994): E24–E31.

treatment plant for Hong Kong, installed the first hazardous waste facility in Indonesia, and become a key investor in Thailand's nascent hazardous waste management industry. WMX is also involved in the establishment of waste-to-energy plants in Taiwan and China, and is looking toward Malaysia for new projects.[32] In addition, WMX invested over U.S.$100 million in a project to build toxic waste dumps and incinerators in Mexico to handle waste from both U.S. and Mexican firms.[33] Because Mexico became a member of the OECD in 1994, the Basel ban on this waste trade does not apply here. This spate of investment by WMX as well as that of other environmental services firms is not surprising. The Asian environmental services market alone has been growing at a rate of 16–20 percent annually and in 1993 was worth U.S.$47 billion, not far behind the U.S.$68 billion value of environmental services in the United States and U.S.$58 billion value in Europe that same year.[34]

While the cleanup market in the developing world has been growing, the foreign firms investing in it have found it frustrating to reap the full benefits because the environmental regulations are as yet still weak compared to those in more industrialized countries. Most governments in the region have been reluctant to force export-oriented industries to incur the high costs associated with private sector treatment of hazardous wastes.[35] On top of the still weak regulations, those that do exist are poorly enforced. This problem appeared to have worsened during the financial crisis that struck that region in 1997–98. A study on industrial pollution in Indonesia after the crisis began in 1997 shows that while industrial output in highly polluting firms surveyed declined by 18 percent, pollution intensity in industrial effluents increased by 15 percent.[36] It is likely that similar results would hold for most hazardous waste–generating industries. So while one might have expected reduced industrial output to result in less waste generation, in fact it has resulted in more. Why? The increase in waste generation during the financial crisis is likely the result of reduced government ability, financially and physically, to monitor and enforce environmental regulations during the time of crisis. Thus, relatively unwatched, those firms still in production are not attempting to comply with

[32] Schwarz 1994, 136; Joshua Karliner, *Toxic Empire* (San Francisco: Political Ecology Group, 1994), 20–21.
[33] Karliner 1994.
[34] Lucas 1995, 32.
[35] Schwarz 1994, 140.
[36] Shakeb Afsah, "Impact of Financial Crisis on Industrial Growth and Environmental Performance in Indonesia," World Bank New Ideas in Pollution Reduction website: www.worldbank.org/nipr/work/shakeb/index.htm.

environmental regulations, presumably because there are short-term cost advantages to not complying. So while the opportunities for cleaning up appear promising, given all of the hazardous waste that has been disposed of improperly in the past, getting those responsible to pay for cleaning up the mess left behind has not always been easy.

To ensure an adequate return, global cleanup firms are increasingly relying on financial support from northern governments and multilateral aid agencies for their investments in the developing world. The World Bank's International Finance Corporation, its private lending arm, has acknowledged that this market is still relatively undeveloped and has offered information to private firms on where to apply for funding for such "environmental projects."[37] Japan, the United States, and other countries, as well as the World Bank and Asian Development Bank, are all providing financial backing to environmental technology firms from more developed countries to allow the cleanup of Asia to be undertaken. The United States Agency for International Development, for example, established a U.S.$117.5 million ten-year U.S.-Asia Environmental Partnership with the explicit aim of linking U.S. environmental technology firms with Asia's needs for less polluting as well as cleanup technologies.[38]

Many governments in the Asia-Pacific region require environment industry firms to come to the bargaining table with financial support in hand.[39] Thus many global cleanup firms make a funding agency their first stop with their investment proposals for waste facilities. This increases their chances both of making a return on the initial investment in what is seen to be a risky business and securing at least some commitment by the host government to attempt to enforce regulations that require the use of the facility.

While these may not be investments in new, state-of-the-art clean technologies, it could be argued that at least the developing world is getting much needed hazardous waste facilities out of these investments, which is better than not having them at all. Old sites that are contaminated with toxic waste do need to be remediated, and wastes that have been piling up waiting for new facilities to treat it do need someplace to go. But so far the

---

[37] International Finance Corporation, *Investing in the Environment: Business Opportunities in Developing Countries* (Washington, D.C.: World Bank, 1992), see especially 25–27.
[38] Lewis Reade, "Enviro-Economics Stirs a Sleeping Giant," *Hazmat World* 7, no. 3 (1994): 32–34; U.S.-Asia Environmental Partnership, ENRIC Activity Profile, at http://www.info.usaid.gov/enric/profiles/499-0015.htm.
[39] Vincent Rocco, "Global Work Requires Rethinking, Says U.S. Cleanup Firm Chief," *Engineering News Record* (May 29, 1995): 47.

focus has been primarily on this type of investment rather than on the transfer of clean technologies that seek to eliminate the generation of such wastes altogether or at least substantially reduce them.

The problem is that truly clean production technology is still facing the problems it faced a decade ago, which are hindering its widespread adoption among global firms. Some estimates show that the international market for these clean technologies is growing, but slowly. Business studies with case examples provide some anecdotal evidence that clean technologies being adopted by some pioneering firms, but there is still a lack of evidence of their systematic adoption, particularly among TNCs operating in the developing world. Meanwhile, it is widely acknowledged that the cleanup industry still makes up the bulk of environmental investment. The transfer primarily of cleanup technologies (as opposed to clean production technologies) tends to perpetuate the problem of double standards by placing emphasis on the continuation of hazardous waste generation rather than on transferring technologies to eliminate or substantially reduce it.

Voluntary Initiatives: The Promises of ISO 14000

If market-based initiatives have been disappointing with respect to promoting clean production, have voluntary codes of conduct for firms fared any better? In recent years several international initiatives have emerged to establish voluntary codes of environmental conduct for industry. These include, for example, the Coalition for Environmentally Responsible Economies' (CERES) Principles, the International Chamber of Commerce's Business Charter for Sustainable Development, and the International Organization for Standardization's ISO 14000 environmental management standards.[40] The idea behind these codes and standards is that firms set their own environmental goals and establish their own procedures to enable them to work toward meeting those goals. Industry has argued that voluntary measures are preferable to command and control regulations set by government because they bring not just environmental benefits but also economic benefits through improved efficiency as well as enhanced public image.[41]

---

[40] For a survey of the measures incorporated in these and other voluntary industry self-regulatory schemes, see UNCTAD, Division on Transnational Corporations and Investment, *Self-Regulation of Environmental Management: An Analysis of Guidelines Set by World Industry Associations for Their Member Firms* (Geneva: United Nations, 1996).
[41] Schmidheiny and WBCSD 1992.

The rise of these voluntary environmental codes at the international level has been partly in response to widespread public perception that industry has played a large role in creating many of today's environmental problems, especially those linked with industrial hazards. Industry also may be looking to these codes as a way to preempt or at least to soften government monitoring and enforcement of environmental regulations. Although these codes ask firms to set their own environmental goals and to commit themselves to preventing pollution, none of them stipulates that firms must meet specific performance or emissions standards.[42] This is where they differ substantially from state-based regulations, which usually are performance based. At the same time, however, these initiatives have increasingly been accepted and to an extent relied on by states and other international bodies as a way to ensure that firms meet environmental goals. This acceptance of voluntary environmental codes by states was outlined in Agenda 21.

The ISO 14000 series of environmental management system (EMS) standards is perhaps one of the more important of the voluntary codes of industry environmental conduct to emerge in recent years. Though the ISO 14000 series was not the first of these private codes to be established, it is rapidly gaining wide recognition and acceptance among businesses and states in rich and poor countries alike. Firms are now seeking to learn more about how to adhere to these new standards, as it is widely seen that these standards will become a de facto business condition for firms that wish to compete in the global marketplace.[43] By the end of 1999, more than thirteen thousand firms in seventy-five countries had obtained certification to the ISO 14001 standard.[44] But will these voluntary environmental management standards help developing countries to avoid problems associated with hazardous waste and to adopt clean production technologies?

The International Organization for Standardization (ISO) has set technical standards for industry since it was established in 1946. The main aim of the ISO is to facilitate international trade and technology transfer by promoting the global compatibility of products and the increased efficiency that comes with technical standardization. The national standards

---

[42] For an overview of these various codes and their contents, see Jennifer Nash and John Ehrenfeld, "Code Green," *Environment* 37, no. 1 (January–February 1996): 16–45; see also UNCTAD 1996.

[43] Charles Denton, "Environmental Management Systems: ISO Standard 14000," *International Environment Reporter* 19, no. 16 (August 7, 1996): 715.

[44] Jason Morrison et al., *Managing a Better Environment: Opportunities and Obstacles for ISO 14001 in Public Policy and Commerce*, Occasional Paper, Executive Summary, Pacific Institute for Studies in Development, Environment and Security, March 2000, 2.

organizations of some 117 countries make up the membership of the ISO. Because some national standards bodies are private institutions while others are public, the ISO is a quasi-private, quasi-public institution. Standards are set by technical committees, subcommittees, and working groups associated with the ISO, which in principle are made up of representatives from industry, research institutes, government authorities, consumer bodies, and ISO representatives.[45]

ISO standards are set through a six-stage process. First is the proposal stage, when the need for a new standard is confirmed. Second is the preparatory stage, when the technical committee or subcommittee prepares a working draft of the standard. Third is the committee stage, when the draft standard is distributed for comments among the member-bodies of the committee. Fourth is the inquiry stage when the draft standard is circulated to all ISO member bodies for comment and voting over a period of five months. Fifth is the approval stage, when all ISO member bodies receive the final draft international standard and vote on it over a period of two months. And sixth is the publication stage, when, once the draft international standard has been approved, it is published as an international standard. Standards are reviewed for confirmation, revision, or withdrawal at least once every five years.[46]

Discussion on establishing the ISO 14000 series of environmental management standards began in the early 1990s when the organizers of the UNCED requested information from the ISO on its activities with respect to the environment. The ISO's response was to establish the Strategic Advisory Group on the Environment (SAGE) to discuss what its role might be in this regard.[47] Talks centered on national and regional EMS standards that were being established in several countries at the time, such as the British Environmental Management Standard BS 7750 and the European Union's Environmental Management and Audit Scheme (EMAS). These initiatives were seen by the advisory group to be potential trade barriers, and it was argued that standardizing environmental management systems across countries was desirable.[48] Following the Earth Summit, SAGE recommended that an international environmental management standard be established by the ISO to help enhance industry's ability to fulfill the role set out for it in Agenda 21.

---

[45] ISO, "Introduction to the ISO," http://www.iso.ch/infoe/intro.html.
[46] See ISO, "Stages of the Development of International Standards," http://www.iso.ch/infoe/proc.html.
[47] ISO, *The ISO 14000 Environment* (ISO Background Document to ISO 14000), March 1996, 2.
[48] Nash and Ehrenfeld 1996, 37.

The ISO focused its efforts on setting up an EMS standard, as opposed to one based on environmental performance or technical specifications, because it judged that the former would be more flexible to situations in different countries, and would be less likely to act as a barrier to trade.[49] It also argued that a standardized management system would be the most effective way of stemming environmental problems at their source.[50]

By early 1993 negotiations on the ISO's international environmental management standards began. These negotiations were overseen by a new technical committee, TC 207, chaired by Canada's standards organization, the Standards Council of Canada. By 1996 standards organizations from forty-nine countries participated actively in TC 207, and there were seventeen observer countries.[51] Also involved in the discussions have been industry groups, government representatives, international organizations and some environmental NGOs, though industry seems to have had a heavier representation at meetings than the others.

The ISO 14000 series covers environmental management standards in six separate areas: environmental management systems, environmental auditing, environmental labeling, environmental performance evaluation, life-cycle assessment, and terms and definitions.[52] Firms can become certified only to the ISO 14001 EMS standard, whereas the other standards serve as guidance documents.[53] As the only certifiable standard and as one of the first of the series to be adopted, the 14001 standard has thus far received the most attention. This standard was published in 1996 and is due for review in 2001. Firms must adhere to certain rules to become certified to this standard, and certification is to be separate for each facility or site. The main criteria for gaining certification to ISO 14001 can be categorized roughly as follows:[54]

- Each facility must have its own environmental policy statement which indicates that it is committed to comply with all applicable

---

[49] ISO, "First Environmental Management System Standards Appear," press release, http://www.iso.ch/presse/PRESSE09.html.
[50] "Environment: ISO 14000 Standards Focus on the User," *ISO Bulletin* 27, no. 2 (February 1996): 6–7.
[51] *The ISO 14000 Environment*, March 1996, p. 8.
[52] Naomi Roht Arriaza, "Shifting the Point of Regulation: The International Organization for Standardization and Global Law-Making on Trade and the Environment," *Ecology Law Quarterly* 22, no. 3 (1995): 502–15.
[53] *ISO Bulletin* (February 1996): 7.
[54] These requirements are outlined in more detail in Benchmark Environmental Consulting, *ISO 14001: An Uncommon Perspective* (Portland, Maine: BEC, 1996), 3.

environmental laws in the jurisdiction in which it is located, and it must be committed to continual improvement and the prevention of pollution.[55]
- The facility must adopt a management system that ensures that it stays in conformance with its own environmental policy statement. Suppliers and contractors are to be encouraged to establish their own EMS that conforms to the ISO 14001 standard.
- The facility must be audited to ensure that the management system is indeed implemented. Firms can self-certify or can opt to be certified by a third-party auditor.
- The firm's environmental policy statement and certification document must be made available to the public upon request.

Auditors must be accredited by national standards-setting bodies, and certified firms are to be reinspected on a regular basis. Certificates are to last for a limited period, generally three years, at which time a full reaudit is required. Policies on reinspection and reaudits are set by national standards-setting bodies and are still being developed.[56]

The standards are designed to promote environmental awareness among firms, and this should provide incentive for environmental improvement. Certified firms are forced at all steps of their management to take environmental considerations into account. The environmental policy developed by firms should spell out clearly that they are committed to comply with all applicable environmental laws in the jurisdiction in which they are located. This should ensure that if the quality of enforcement is poor, firms will still abide by local regulations. The commitment to continual improvement in environmental matters and the prevention of pollution is supposed to encourage firms to adopt cleaner technologies rather than focusing on cleaning up after the fact. The audit of the firm's environmental policy document and management practices should provide a somewhat objective analysis as to whether the firm in question has indeed met these criteria. In requesting firms to encourage suppliers to follow the standard, good environmental management practices should spread up and down the supply chain, as well as across borders. The availability of the certification document to those requesting to see it should provide assurance that the entire process is legitimate and above board.

---

[55] Joseph Cascio, "The Increasing Importance of International Standards to the US Industry Community and the Impact of ISO 14000," *EM* (November 1996): 18.
[56] See Riva Krut and Harris Gleckman, *ISO 14001: A Missed Opportunity for Global Sustainable Industrial Development* (London: Earthscan, 1998), 16.

ISO 14000, Hazardous Waste, and Clean Production

Although the expectations from the standard are reasonable, concerns have been expressed about its capacity actually to improve environmental performance. Critics claim that the standards as they are structured do not do much to reduce hazardous waste generation or to facilitate cleaner production technology transfer to developing countries. Rather, it appears as though at best the standards will maintain the status quo on these issues, and at worst they might actually make these problems even more pronounced.

One of the main criticisms made of the ISO 14001 standard is that it is based on environmental management and not on environmental performance.[57] Firms are not required to reduce their generation of hazardous waste or to report pollution emissions levels. Nor are there any stipulations in the standards calling for the transfer of cleaner technologies to developing countries. Firms must only ensure that management systems are dedicated to meet existing environmental laws in the country in which they are operating and that they are committed to "continual improvement" and "prevention of pollution." Critics have charged that this latter concept, the prevention of pollution is especially misleading because the ISO's definition of prevention of pollution includes end of the pipe type measures that are only remedial, whereas pollution prevention does not.[58] The standards thus provide little incentive to reduce hazardous waste. Because the standards require firms to comply with existing laws only in the country in which they are operating, the ISO 14000 standards may be very costly to implement but may not actually make much difference in environmental quality, especially in developing countries. Indeed, firms that already have a strong EMS in place have found the ISO 14000 standards to be weak at best.[59]

Because firms set their own environmental policy and goals under the standard and are judged only against their own management system when they are audited, they can set very low goals and still become certified to

---

[57] For an important critique, see BEC 1996; see also Roht-Arriaza 1995; Gareth Porter, "Little Effect on Environmental Performance," *Environmental Forum* 12, no. 6 (November–December 1995): 43–44.

[58] "ISO 14001 Should Not Be Required by Law or Regulation, Attorney Says," *International Environment Reporter* 19, no. 4 (February 21, 1996): 122.

[59] Ann Thayer, "Chemical Companies Take Wait and See Stance toward ISO 14000 Standards," *Chemical and Engineering News* 74, no. 14 (April 1, 1996): 11; Marlon Allen, "Despite Feeling Their EMS Plans are Better, Most Global Firms Creating ISO 14001 Strategy," *International Environment Reporter* 20, no. 24 (November 26, 1997): 1088–89.

the ISO 14001 standard. At one information seminar in Thailand, for example, a Western consultant advised Asian firms not to set their goals too high when formulating their own EMS and to aim only for compliance with existing national environmental regulations.[60] Indeed, because the 14001 standard is not performance based, firms in developing countries, either locally based or TNC subsidiaries, have little incentive to go beyond meeting the existing environmental laws. This approach of the ISO 14000 standards is very different from the recommendation in *Agenda 21* that TNCs apply the same standards regarding hazardous waste disposal in their operations worldwide. Instead, it will likely only ensure that differences in standards between different countries will remain that way and may encourage a continuation of double standards in firms' behavior in their operations in developed and developing countries.

The 14001 standard provides little incentive for firms to adopt cleaner technologies, and TNCs are not required to transfer such technologies to their subsidiaries in developing countries. There was debate at the drafting stage of the standard over whether it should require firms to adopt the best available technologies in order to gain certification, a suggestion made by European participants. In the end, this requirement was not incorporated, and the draft standard only encouraged firms to consider implementing the best available technology where it was "appropriate and economically viable."[61] Joseph Cascio, a key proponent of the standards who was involved in the drafting process, has since remarked that the goal of the standards is not to get every country to use the same best technologies, reasoning that, "if all countries had to use the best available technology, what was supposed to be a management standard would turn into a performance standard, making the standard unworkable and unattainable for many developing countries."[62] Because so few developing countries were involved in drafting the standards, it is not clear that they were the main opponents of the requirement of best available technology. The United States in fact was very much opposed to the idea because it was worried about the legal implications.[63] In the end, the standards show little commitment to transferring environmentally sound technologies to developing countries.

---

[60] "Foreign Firms Expected to Lead the Way on ISO 14000 Certification in Thailand," *International Environment Reporter* 18, no. 27 (November 1, 1995): 84.
[61] Quoted in Roht-Arriaza 1995, 506.
[62] Quoted in "ISO 14000 Standards Tailored to Meet the Needs of All Countries," *Quality Progress* 28 (June 1995): 21.
[63] Roht Arriaza 1995, 506.

There is also some concern that adherence to the ISO 14000 standards will not strengthen and may in fact weaken regulatory frameworks at the national and international levels because the standards are based on firms' self-set goals and are strictly voluntary. There may be reason to suspect that existing environmental laws will be watered down in developing countries if the ISO 14000 standards become widely used by firms. Cascio has argued that the standards may lead some countries to discover that they have more laws on their books than they can ever enforce, given their resources. He sees this as providing the impetus for some developing country governments to redraft their environmental laws so that they can meet existing resources and capabilities.[64] Although this may increase the ability of firms to comply with legal requirements, it could actually weaken the existing regulatory framework in those countries.

An additional reason why the regulatory framework in developing countries may not improve is that with the legitimacy given to the standards in the 1994 GATT agreement, it is almost certain that developing countries will not establish national EMS standards that are more stringent than the ISO standards because they may fear that the standards could be challenged as trade barriers. The Technical Barriers to Trade (TBT) agreement under the 1994 GATT indicated that internationally set standards that are either in existence or imminent are to be followed by GATT member countries as a way to reduce technical barriers to trade.[65] The GATT description of standards setting does not include standards established intergovernmentally or by UN bodies. Rather, governmental standards and regulations are seen to be "technical regulations" which the TBT agreement sees as creating potential trade barriers. Because the ISO 14000 standards were being drafted at the time, they were seen as imminent, and are recognized as international standards under the GATT. It will thus be possible for states to challenge other states' national environmental management standards under the GATT as a technical barrier to trade if they do not conform to ISO standards.[66] Such a challenge is most likely to happen only if a country pursues standards that are more stringent than the ISO standards. Thus the ISO 14000 series has the effect of being a ceiling for international EMS standards.

---

[64] Cascio 1996, 18.
[65] BEC 1996, 4–5; Matthias Finger and Ludivine Tamiotti, "The Emerging Linkage between the WTO and the ISO: Implications for Developing Countries," mimeo, 1999, 4.
[66] Krut and Gleckman 1998, 68–70.

TNCs as well as local firms in developing countries may also try to use ISO 14000 certification to obtain regulatory relief, which firms are already trying to do in the United States and other industrialized countries.[67] There are signs that firms in developing countries are already pushing for regulatory relief, such as reduced monitoring, in return for conformance with the ISO 14000 standards. In Argentina, for example, firms are pressuring the government to relax environmental regulations for those firms that have obtained certification to the ISO 14001 standard. The Mexican government, currently overhauling its domestic environmental regulations, is considering some form of regulatory relief along similar lines. Governments are responding with a variety of measures that take ISO 14001 certification into account in the monitoring and enforcement of regulations.[68] Although these developments may make it easier for firms to comply with legal requirements in the country in which they operate, the existing regulatory framework in those countries could be weakened if they cease to monitor and enforce regulations for companies that are ISO 14001 certified.

A further problem that some critics of the standards have pointed out is that the ISO 14000 standards do not even mention existing international environmental treaties such as the Basel Convention as a concern for firms.[69] This is seen to be a problematic issue because firms are not required to be aware of or to comply with these internationally agreed environmental rules. The failure to recognize the importance of firms' adherence to principles and stipulations set out in international environmental agreements goes against the spirit of Agenda 21, which stressed their importance.

The Privatization of Global Environmental Governance

Concerns have been raised not just about the environmental impact of the ISO 14000 standards but also about the standard-setting process. The nature of the standards and their growing acceptance by governments and intergovernmental bodies, take the ISO into the realm of public policy

---

[67] See, for example, Ronald Begley, "ISO 14000: A Step toward Industry Self-Regulation," *Environmental Science and Technology News* 30, no. 7 (1996): 298A; Rick Mullin and Kara Sissell, "Merging Business and Environment," *Chemical Week* 158, no. 38 (October 9, 1996): 52–53.

[68] Lawrence Speer, "From Command-and-Control to Self-Regulation: The Role of Environmental Management Systems," *International Environment Reporter* 20, no. 5 (March 5, 1997): 227–28; Finger and Tamiotti 1999, 9.

[69] Krut and Gleckman 1998, 36–39.

rather than just the setting of technical standards. But the participation processes of the ISO have not changed to take this into account. Developing country governments and environmental NGOs in particular were largely absent from the standard-setting process, in which representatives from industrialized countries and TNCs were dominant. The official delegates to the TC 207 meetings are the representatives of the member bodies, but these participants come from various sectors, depending on whether the member body is a public or a private organization.

Developing countries were largely unrepresented during much of the negotiation of the standards.[70] The low level of developing country participation is in part because these countries have a much lower representation in the ISO generally, and they were conspicuously underrepresented in TC 207 in the early stages of setting the ISO 14000 standards in particular. Only two developing countries (Cuba and South Africa) had representatives at the first TC 207 meeting in early 1993. When the ISO 14001 and 14004 standards were voted on and approved as draft international standards in Oslo in June 1995, the standards bodies from only six developing countries had actively participated in the process up to that time, and these were mainly rapidly industrializing developing countries. Thus while 92 percent of industrialized countries were present at the Oslo meeting and voted on the standards, only 16 percent of developing countries were present and voting.[71]

In the negotiation of UN-based environmental treaties, developing countries have generally pushed for phase-in periods and economic assistance to enable them to meet the treaties' demands without incurring significant costs that may hinder economic growth. Developing countries might not have pushed for stricter EMS standards if they had been more active in the earlier stages. But representatives from developing countries present at the Oslo meeting expressed their disappointment at the lack of attention in the discussions to issues of technology transfer, a phase-in period for developing countries, and commitment to equal representation from developing countries in the standard-setting process.[72]

One of the key reasons for the lack of participation of developing country standards bodies and other representatives is that the cost of attending technical committee and plenary ISO meetings is borne by each

---

[70] Roht-Arriaza 1995, 528.
[71] Krut and Gleckman 1998, 42.
[72] UNCTAD, *ISO 14001: International Environmental Management Systems Standards: Five Key Questions for Developing Country Officials*, draft report (Geneva: United Nations, 1996), 22–38.

participant. These costs can be substantial, particularly for representatives from developing countries. At the June 1995 Oslo meeting, there was an increased number of developing country representatives who had joined the TC 207, mainly as a result of financial assistance from the Netherlands and Finland, which was channeled through the Developing Countries (DEVCO) assistance program organized by the ISO. This program is intended to promote environmental management and ISO 14000 in the developing world by raising funds from donors to support developing country representation at TC 207 meetings. Some twenty-two representatives from the developing world were sponsored under this program in 1995 and twenty-three in 1996. The program funds only two representatives per country, generally one from a standards-setting organization, if it exists, and one from a government environmental agency or environmental NGO.[73] Thus developing countries had much smaller delegations than those of industrialized countries, making participation in the twenty-five working groups and subcommittees of TC 207 particularly difficult.

Very few nongovernmental organizations were involved in drafting the standards. Organizations can apply to be liaison members of the ISO, which entitles them to receive materials and participate but not vote in discussions in the standards-setting process. NGOs, like developing countries, face the constraint of having to pay their own costs of attendance at meetings. Of the twenty-two liaison members that were involved in drafting the ISO 14001 standard, only eight were environmental NGOs, while the rest were industry groups and international organizations.[74] NGOs may be invited to participate on the delegations of national standards-setting bodies, but few NGOs have been asked to do so. The World Wide Fund for Nature has criticized the ISO for its failure to involve more developing country standards organizations and environmental NGOs in the ISO 14000 standards-setting process. It also complained that the organization suffers from a serious lack of transparency, citing the ISO's adoption in June 1996 of a policy effectively banning media access to its meetings.[75] This exclusion of the media and lack of NGO participation is very different from UN-sponsored environmental treaty negotiations such as

---

[73] Interview with Anwar El Tawil, ISO Developing Countries Program, Geneva, July 9, 1996.
[74] International Institute for Sustainable Development, *Global Green Standards: ISO 14000 and Sustainable Development* (Winnipeg: IISD, 1996), 93.
[75] "WWF Calls on ISO to Clear Up Confusion Surrounding Extension of ISO 14001 to Forestry," *International Environment Reporter* 19, no. 19 (September 18, 1996): 811.

the Basel Convention. If more environmental NGOs had participated, it is likely that they would have pressed for more performance-based indicators in the standards.

While developing countries and NGOs had little input into the standard-setting process, industry groups and environmental consultants were heavily represented in the activities of TC 207.[76] About four hundred representatives from U.S. industries (including chemical, petroleum, electronics, and consulting firms) were actively involved, while only about twenty representatives of government and public interest groups were involved.[77] The committee that drafted the standards was made up mostly of executives from large multinational firms, standards-setting bodies, and consulting firms. The chairs of the subcommittees and working groups of TC 207 all came from industrialized countries; half of the working group chairs were employed by major multinational corporations.[78]

Though the participation in the drafting of the standards was limited, their impact has been much wider than just on private industry firms that choose to adopt them. Many governments and intergovernmental organizations such as the WTO and the World Bank have recognized and endorsed the standards. The most important of these is the WTO recognition of the standards as international standards under international trade law. In this way, standards that were negotiated and drafted mainly by private industry actors and designed to be voluntary have taken on enormous public and environmental significance.

Conclusion

Following UNCED there was great hope that poor countries could avoid the rich countries' mistakes with respect to hazardous waste generation. TNCs were seen as a key part of this process. They were to play a central role in the transfer of clean production technologies through the market, and through voluntary environmental management initiatives they would diffuse an environmental ethic into corporations around the world. In the years since UNCED, however, these developments have led to some disappointment regarding the goals of hazardous waste reduction and clean technology transfer. The push for green investment in environmental technologies has thus far been dominated by cleanup technologies rather than

---

[76] BEC 1996, 12.
[77] Nash and Ehrenfeld 1996, 37.
[78] Krut and Gleckman 1998, 54–55.

truly clean ones. This practice encourages double standards and a continuation of hazardous waste generation rather than its prevention.

Voluntary EMS initiatives such as the ISO 14000 offer much promise for improving firms' environmental performance. But they do little to ensure that firms have a strong incentive actually to reduce hazardous waste generation and technology transfer. It is too early to assess whether the standards have in practice improved firms' environmental performance, though some claim that ISO-certified firms are changing their mentalities with respect to the environment.[79] Others stress their potential to bring such change but also highlight the pitfalls.[80] The process by which the standards were set also represents a privatization of global environmental governance, whereby private industry actors are determining standards that have taken on international public significance, affecting not just industry but far beyond.

The activity of industry in the area of market-based and voluntary initiatives for the greening of business represents the attempt by private actors to capture the regulatory process in ways that open the door for a continuation of hazard transfer. Though environmental NGOs are vocal in calling for the adoption of clean production technologies on a global scale, they have much less voice in the global dialogue regarding this issue than they do in the negotiation of multilateral environmental agreements. Industry has taken a lead role in setting the rules that it finds easiest to comply with. Governments, especially those in the industrialized world, have welcomed and supported these moves on the part of industry. At the same time, the global economy has facilitated these developments. The phenomenon of economic globalization is used by some governments as a key reason for letting industry self-regulate because it is argued that regulation becomes more difficult as the global economy becomes more integrated. But the move in the direction of cleanup as opposed to clean production technologies and voluntary initiatives such as the ISO 14000 have done little to address the problems of toxic waste generation and management in developing countries, for both TNCs and local firms.

Meanwhile, developing countries and in particular environmental NGOs seem to have been shut out of the process of developing voluntary standards for industry. Indeed, the development of the ISO standards was almost the direct opposite of the Basel negotiations, where environmental

---

[79] "Companies Are Benefiting from ISO 14000, Reports the Times," *ISO 9000 +ISO 14000 News* 7, no. 3 (1998): 25–29.
[80] Morrison et al. 2000.

NGOs and developing countries had the upper hand. Criticisms of the ISO 14000 standard-setting process have emerged, but this response came somewhat late in the day, once the 14001 standard had already been adopted. Because the standard is due for review and possible revision in 2001, there is some scope to incorporate NGO and developing country suggestions for changes. This would first require measures to increase their participation in the process.

7
Conclusion: Prospects for Clean Production on a Global Scale

Hazards have made their way from rich to poor countries through a variety of outlets. New regulations to stop one form of hazard transfer are soon circumvented either by the opening up of new channels through which hazards can be relocated or through a rewriting of the rules. As a result, the problem has evolved from one that was focused primarily on the export of hazardous waste from rich to poor countries to a much broader practice. This broader understanding of hazard transfer also encompasses the export of waste for recovery; attempts by industry to weaken international legal instruments by various means; the transfer of hazardous industries and production processes; the export of cleanup as opposed to clean technologies; and the writing of rules to encourage clean production among firms in ways that allow for a continuation of hazard transfer. The problem is thus a dynamic and multifaceted one, constantly changing in form as new rules are put in place.

Nonstate actors have played a crucial role in the dynamic nature of hazard transfer. Transnational corporate actors have been principal players in the initial movement of hazards. Environmental nongovernmental organizations have been key in pushing for stronger global rules to stop hazard transfer. And corporate lobby groups have been active in trying to ensure that the global rules that are adopted do not harm the economic interests of their member firms. It seems that every time new rules are agreed or strengthened, new outlets for hazard transfer arise. The dynamic nature of the global economy itself has facilitated this expansion of the problem, and nonstate corporate actors, corporations as well as industry lobby groups, have actively sought out these new outlets.

Most conventional accounts of global environmental conflict and cooperation fail to identify the dynamic nature of global environmental problems and their relationship to economic globalization. Mainstream accounts also tend to focus on states as the primary actors in the making of global environmental rules. This book has shown that global environmental problems such as hazard transfer are dynamic in nature, shaped by the global economy and inequality. It has also shown that nonstate actors were key in the evolution of those problems and the global agreements that attempt to mitigate their impact. I hope that this book has contributed to a broader understanding of the nature of global environmental problems in an age of economic globalization and that it will assist in identifying ways to improve international policy initiatives aimed to address them.

The transfer of hazards in these various forms is still current and continues to evolve. Under these circumstances, it is vital to continue to keep a close watch on both old and new trends. In light of these developments, I outline a provisional list of the measures that I see as necessary to tackle the hazard transfer problem. I argue that action on a number of fronts, with measures that are flexible and dynamic, will be key in efforts to address the problem.

Current Trends in Hazard Transfer

With the adoption of the Ban Decision in 1994 and the Basel Ban Amendment in 1995, the number of cases of waste exports from rich to poor countries has diminished dramatically. In the late 1980s and early 1990s environmental groups such as Greenpeace International were publishing databases with details from hundreds of rich to poor country waste trade cases. Even though the ban amendment has not been ratified and is thus not officially in force, there has been a de facto ban on such transfers with a number of national and regional import and export bans. Today the number of waste transfers from rich to poor countries has dwindled to just a few per year. Environmental groups such as BAN see this as a sign that action on the part of the international community to put an end to the waste trade has been successful. Without these efforts, things might have been far worse. It is in this context that environmental NGOs tracking hazard transfer have continued to publicize ongoing cases of waste exports from rich to poor countries. These cases remind us that the problem continues to evolve, and that in the absence of pressure for a legally binding ban it could resurface in full force. There are several factors that help to explain this continuation of the trade. First, not all countries are party to the Basel Convention, and many parties have yet to ratify the ban amendment. Second, the problem of illegal transfers of hazardous waste has not been fully eliminated. Third, developing countries have little legal protection from hazard transfer from other developing countries. Hazard transfers continue between countries at different relative economic positions.

Several recent cases illustrate that hazardous waste transfer between countries in the OECD and non-OECD countries has continued to be a problem, even for countries that have ratified the Basel Convention. In late 1999, the government of the Philippines announced that a Japanese firm had shipped 2,700 metric tons of wastes for disposal in Manila, in direct contravention of the Basel Convention. The wastes were labeled as paper for recycling, but in fact they contained a mix of hazardous medical and industrial wastes that were not suitable for recycling.[1] The Philippine government discovered the wastes when they were abandoned at Manila Harbor. The Japanese government was clearly embarrassed by the incident and removed the waste from the Philippines within thirty days, as required by the Basel Convention. One report speculated that the exports may have been spurred by a tightening of Japanese laws for dioxin emissions which forced the firm to shut down its incinerators.[2] The case thus also illustrates how more stringent regulations in richer countries can contribute to hazard transfer. The firm responsible was also being investigated for illegal waste dumping within Japan. The president of the firm went into hiding when authorities in Japan sought his arrest, and in July

[1] "Illegal Dumping," *Mainichi Daily News*, Niigata, Japan, January 13, 2000.
[2] "Japan: Philippines Case Bares Inadequacy of Waste Rules," *Yomiuri Shimbun*, Tokyo, Japan, January 12, 2000.

2000 it was discovered that he had committed suicide.[3] Another company official involved in the scandal who had fled to Cambodia was recently arrested and extradited back to Japan.[4]

This incident did not prevent others from contravening the convention, however. In September 2000, South Africa authorized the import from an Australian firm of sixty tons of waste that contained levels of lead and arsenic considered hazardous. In theory, the waste was to be used for recycling research, and remaining hazardous residues were to be shipped back to Australia. Environmental NGOs have denounced this agreement between the countries. According to Jim Puckett of BAN, "This is the first time that the 1994 Basel Convention's Dumping Ban decision has been intentionally violated."[5] Neither country has yet ratified the ban amendment, though in principle parties to the convention, which include both South Africa and Australia, have agreed to abide by decisions taken by the conference of the parties.

India has also continued to import hazardous and potentially hazardous wastes for recycling purposes. According to recent Indian government data reprinted by Greenpeace, recycling firms in India have imported over one hundred thousand metric tons of wastes such as used batteries, used lubricating oils, zinc ash and residue, copper cables potentially coated with PVC, and toxic metal scrap from March 1998 to March 1999. These wastes came not only from industrialized countries that are members of the OECD such as Australia, Canada, Germany, the United States and the United Kingdom but also from other developing countries such as Hong Kong, Malaysia, the Philippines, Singapore, Saudi Arabia, and the United Arab Emirates.[6] Though India is a party to the Basel Convention and its national law bans the import of toxic waste, it has not ratified the ban amendment and continues to turn a blind eye to imports of certain hazardous wastes. Nonetheless, Greenpeace alleges that many of these imports are illegal under the Basel Convention and Indian law. India, however, did recently lift a ban on the import of zinc ash scrap and is considering allowing imports of lead scrap.[7]

---

[3] "Japan: Waste Company Boss Suicides with Poison," *Mainchi Daily News*, Niigata, Japan, July 23, 2000.
[4] "Japanese Waste Trader Arrested," *Haznews*, no. 147 (September 2000).
[5] BAN, "Environmentalists Call on South Africa to Reject Imports of Hazardous Waste," press release, September 14, 2000. www.ban.org.
[6] Greenpeace, "Toxic Waste—Poisons from the Industrialised World," www.ban.org.
[7] Camila Reed, "India Pushes for Changes on Basel Waste Trade Rule," Reuters (December 17, 1999). Posted on the Basel Action Network website: www.ban.org/ban_news/india_pushes.html.

The import to India of hazardous wastes from other developing countries highlights a troubling trend of hazard transfer between developing countries at different income levels. Much of the attention on the trade has been focused on the export from the rich industrialized countries that are members of the OECD to non-OECD countries. But with the rapid industrialization in the past few decades of certain developing countries in East and Southeast Asia, the rich-poor divide among countries that are not members of the OECD has widened. Along with this growing income gap has come the problem of cross-border hazard transfers among developing countries.

India is not the only place where this problem occurred. Another high-profile example is the export of three thousand metric tons of mercury-contaminated industrial waste from Taiwan to Cambodia in December 1998. In this case, a Taiwanese firm, Formosa Plastics Group (FPG), was unable to obtain permission to dump the mercury-contaminated waste locally so it hired a waste brokerage firm to get rid of it. The waste broker exported it to the port town of Sihanoukville in Cambodia, without a permit. Neither country is a party to the Basel Convention. Typical of such transfers, the waste was labeled as another product, in this case "polyester chip," and the shipping papers called it "cement cake." The bags that the waste was shipped in were highly sought after by local residents and were looted when the waste was dumped on vacant land near Sihanoukville. The Taiwanese government noticed that the waste was missing, but it was too late to stop the damage. Several deaths have been attributed to the waste, two from contact with it and the bags it was shipped in and four from a stampede out of the town that occurred when news hit that toxic waste from Taiwan had been dumped there.[8]

This incident received international attention when environmental NGOs publicized it widely. FPG finally agreed to remove the waste from Cambodia, but it did not immediately plan to dispose of it at home.[9] Instead, it tried to export the waste to various locations in the United States, France, and Germany. These attempts were unsuccessful because BAN-affiliated NGOs in those countries organized protests against its import. Finally, in mid-2000, FPG agreed to reimport the waste and dispose of it in Taiwan.[10]

---

[8] Lawrence Speer, "Environmentalists Assail Taiwan's Plans to Ship Waste to French Treatment Facility," *International Environment Reporter* 22, no. 21 (October 13, 1999): 830.
[9] Glen Perkinson, "Company Agrees to Take Back Mercury-Laden Waste Sent to Cambodia," *International Environment Reporter* 22, no. 5 (March 3, 1999): 204.
[10] For a full account of the incident, see BAN, "Victory for Global Environmental Justice: Toxic Waste Dumped on Cambodia Will Finally Be Treated by Producer," press release, Seattle, June 9, 2000, www.ban.org.

These cases illustrate that despite a global consensus in principle on the need to ban waste exports to poor countries, and despite a significant decline in cases of hazardous waste transfers from rich to poor countries, in practice some developing countries still have been recipients of wastes from more industrialized countries, both in the OECD and the non-OECD categories. Neither Taiwan nor Cambodia is a party to the Basel Convention. Even if they were, because neither country is a member of the OECD, the shipment would not be banned by the convention. Developing countries are still vulnerable, either because they are misled about the contents of imports or because they are left with little choice—they need the funds for providing disposal services or they feel that they need the raw materials to be gained from recycling operations. Some developing countries, such as India, may well be ambiguous about the ban.[11]

Can Hazard Transfer Be Halted?

If hazard transfer is a constantly evolving problem, what can be done at the global level to halt it? This is an important question that is no doubt deeply pondered by those who are committed to end this phenomenon. The ultimate solution to the hazard transfer problem is fairly easy to identify. If hazardous wastes are not generated in the first place, they cannot be transferred. The reduction and eventual elimination of hazardous waste generation on a global scale is thus the ultimate goal. Getting there, however, will take a great deal of effort and political will. I explained in Chapter 6 some of the obstacles to eliminating waste generation. These include the greater profitability of cleanup as opposed to clean technologies and the capture by industry players of the process for developing industry guidelines for cleaner production with the result that they do not provide solid enough incentives for industry to abandon hazardous production processes. Though there have been roadblocks to the widespread adoption of clean production in the past, there are ways to encourage its adoption in the future.

The Basel Convention has from its inception had waste minimization as one of its principal goals. Efforts during the first decade of the convention, however, focused almost exclusively on controlling transfrontier movements of waste. The concept of hazardous waste minimization was

---

[11] Katharina Kummer, "Hazardous Waste: Accepted and Hidden Realities of the Basel Ban on Hazardous Waste Exports," *International Environment Reporter* 23, no. 21 (October 11, 2000): 808.

recently reaffirmed by the conference of parties to the Basel Convention. At COP-5, the parties adopted the Basel Declaration on Environmentally Sound Management to guide the convention over the next decade, giving priority to waste minimization. This move brings the Basel process fully into the politics of clean production, a move welcomed by many. But given the dynamic nature of the problem, the fact that the environmentally sound management (ESM) provisions of the convention are not legally binding, and its susceptibility to influence by industry lobby groups, relying solely on the Basel Convention is not likely to be entirely effective or satisfactory at achieving a global adoption of clean production. Strengthening the convention via amendment would help to remedy this situation. But other measures will also be necessary.

For a global environmental problem that is dynamic and broad, solutions must also possess these features. There is no one magic pill that can address the problem, but rather an array of actions are warranted. Below I outline the main elements that I see as crucial for beginning to put an end to the movement of hazards from rich to poor countries. The ultimate aim of these measures is to close the cheap and dirty options for firms and provide incentives for adopting cleaner methods and transferring clean production technologies to developing countries. They include both carrots and sticks and involve action by governments, global institutions, NGOs, and TNCs themselves. Some of these suggestions are well accepted, and work has already begun on them through the Basel Convention and other organizations. Others are more controversial and will require more political will than has been mustered thus far.

Commitment to Strengthening the Basel Convention

It is extremely important that governments give stronger commitment to strengthening and abiding by the Basel Convention than they currently do. Adoption of the convention and the ban amendment, implementation of its provisions, and strengthening the recently adopted liability protocol are all vital for the promotion of cleaner production. Closing off the options for the export of hazardous wastes to poorer countries will be necessary before firms around the world begin to adopt clean production practices.

It has been over a decade since the agreement was first adopted and over five years since the ban amendment was adopted. By April 2001 there were 146 parties to the convention. But there are some notable exceptions, and in total around a quarter of countries in the global community are not yet parties. The United States, for example, the world's largest generator and a major exporter of hazardous wastes, has signed but has not ratified the convention. Several attempts have been made in the United States since 1991 to pass legislation to enable ratification, but these have not been successful. Since the waste lists (Annexes VIII and IX) were adopted at COP-4 in early 1998, the U.S. administration and U.S. industry have been in favor of ratification of the convention. But industry has opposed the ratification of the ban amendment, while the administration was less sure. There was an expectation that implementing legislation would be passed in 1999, which would have paved the way for ratification in time for the United States to become a party to the convention before COP-5. But this did not in fact occur. Mostafa Tolba lamented in his speech to delegates at that meeting, "I fail to find an explanation for this when the U.S. delegation over the eighteen months of negotiations regularly pressed for assurance that the convention provisions are not inconsistent with U.S. national laws and regulations."[12] The United States is still committed to ratify the convention, according to a State Department official, but technical and legal issues have complicated the process.[13]

The Basel Ban Amendment has been slower to come into force than the convention itself. However, it must be understood that only twenty ratifications were required for the Basel Convention to come into force, while sixty-two ratifications are required for the Ban Amendment. As of April 2001, twenty-four parties had ratified the amendment. This number falls far short of the sixty-two ratifications necessary. Some experts, such as Katharina Kummer, have argued that this is extremely slow, indicating a lack of real commitment to the principles in the ban especially among developing countries.[14] But environmental NGOs stress that the pace is not necessarily any slower than with other global environmental agreements, particularly amendments. According to Jim Puckett of BAN:

> Compared with other amendments in international law, the Ban Amendment can hardly be said to be stalled or unduly slow in coming into force. It's on a steady pace and these things do take time. Most significant really is that already the 15 countries of the European Union and the 3 member states of the European Free Trade

---

[12] Mostafa K. Tolba, former executive director of UNEP and guest of honour at the ministerial segment of the fifth Conference of the Parties to the Basel Convention, speech, Basel, Switzerland, December 9, 1999, 3.
[13] Daniel Pruzin, "Former UNEP Chief Blasts United States for Failing to Ratify 10-Year-Old Pact," *International Environment Reporter* 22, no. 25 (December 8, 1999): 975.
[14] See, for example, Kummer 2000.

Association countries have implemented the ban, which means that for more than half of the OECD countries to which the export ban applies, the ban is law.[15]

Whether the process is unduly slow or not, a stronger commitment to the amendment does need to be made via further ratifications to bring it into force. The solidarity of developing countries and the strong relationship between environmental NGOs and developing countries that was apparent in the late 1980s and early 1990s may be weakened by further delay in the ratification of the ban amendment. Several developing countries, including India, Senegal, Brazil, and the Philippines, have been courted by industry groups in an attempt to win them over to their point of view. This strategy may be working. India, for example, broke ranks with the majority of developing countries and called for a review of the Basel Convention's rules on exports at COP-5.[16] Its position stems from a concern over a shortage of lead and zinc in that country, metals which it wishes to import in scrap form.

Stronger commitment is also needed in implementing the convention among the parties. Many parties have failed to fulfill obligations as set out in the treaty. A particular problem has been reporting of data, a key mechanism by which implementation and compliance with the convention are monitored. Very few parties have reported information on hazardous wastes generation and trade to the convention secretariat as required by Articles 13 and 16. In 1996, for example, only twenty-six parties responded to the secretariat's annual questionnaire requesting these data, though this number increased to sixty-three parties in 1997, still less than half of the parties.[17] The lack of adequate information and statistics has made it extremely difficult to assess whether the convention is making a strong impact on reducing the generation and trade of toxic waste, and even secretariat documents note that the data it presents should be used with extreme caution.

Illegal transfers of waste have continued, indicating a further lack of commitment to the convention. Waste transfers are considered in contravention

---

[15] Jim Puckett, personal communication, October 16, 2000.
[16] Reed 1999.
[17] Secretariat of the Basel Convention, "Generation and Transboundary Movements of Hazardous Wastes and Other Wastes: 1996 Statistics," Basel Convention Series/SBC No: 99/006, Geneva, June 1999; Secretariat of the Basel Convention, "Reporting and Transmission of Information under the Basel Convention for the Year 1997," Basel Convention Series/SBC No: 99/011, Geneva, October 1999.

of the convention if they lack proper notification and consent, if the wastes do not conform with the shipping documents, or if they are shipped in a deliberate attempt to defy the rules of the convention. Parties are also asked to implement domestic legislation to punish illegal waste traffickers.[18] The secretariat of the convention has cooperated in recent years with INTERPOL and the World Customs Organization to improve detection of illegal transfers. But these transfers have continued, as several of the examples cited in the first part of this chapter indicate. It is difficult to know the full extent of the illegal traffic in hazardous waste, as it is obvious that those involved attempt to avoid detection.[19] The Technical Working Group and the Legal Working Group of the Basel Convention are currently working on developing procedures to help prevent, identify, monitor, and manage illegal transfers.[20] It is important that these procedures come into place soon.

In an attempt to improve implementation and compliance, the parties requested the Legal Working Group to begin work on establishing a mechanism to promote implementation of and compliance with the convention. This draft document is scheduled to be presented at COP-6.[21] Whether this proposed mechanism will be successful remains to be seen.

Commitment to ratifying and strengthening the recently adopted Basel Protocol on Liability and Compensation will also be important. Negotiations on this protocol have been ongoing since 1993. They were completed in late 1999 and adopted at COP-5. A main impetus for the development of this protocol was the concern of developing countries that they lacked the funds and technology to deal with illegal imports of hazardous waste and their impact on their local environments. The protocol thus provides rules for compensating countries that are victims of environmental damage caused by waste imports and determines liable parties that must provide that compensation.

Agreement on the protocol was difficult to reach. Negotiations continued right up until the last moment. Two key controversial issues that

---

[18] *The Basel Convention*, Article 9.
[19] For a discussion of the problem of illegal transfers of waste and the Basel Convention, see Jonathan Krueger, *International Trade and the Basel Convention* (London: Earthscan, 1999), 87–95.
[20] Secretariat of the Basel Convention, *Report of the Fifth Meeting of the Conference of the Parties to the Basel Convention*, UNEP/CHW.5/29, December 10, 1999, 49.
[21] Secretariat of the Basel Convention, "Monitoring the Implementation of and Compliance with the Obligations Set out by the Basel Convention," Decision V/16, *Report of the Fifth Meeting of the Conference of the Parties to the Basel Convention*, UNEP/CHW.5/29, December 10, 1999, 44–46.

dominated the negotiations in the last months before its adoption centered on whether an "opt out" clause would be included and whether contributions to the proposed compensation fund would be mandatory. A split in positions between OECD and non-OECD countries emerged on these issues, reminiscent of the earlier negotiations on the original text of the Basel Convention and the negotiation of the Basel Ban. Despite the differences between these countries, a push was made to adopt the protocol as a symbolic gesture at COP-5, the tenth anniversary meeting of the parties. In the end, the view of OECD countries prevailed, and the protocol includes provisions allowing for OECD countries to opt out of the protocol so long as they are party to a multilateral or regional agreement under Article 11 of the Basel Convention which meets the same objectives as the protocol. The compensation fund attached to the protocol was made voluntary rather than mandatory. Developing countries were troubled by the outcome, arguing that if rich countries could opt out of the protocol, they would be unlikely to contribute to the compensation fund.[22]

Environmental NGOs also condemned the protocol. In addition to citing the concerns of the developing countries regarding the opt out clause and the compensation fund, environmental NGOs pointed out other flaws in the agreement. These include the failure to assign liability to the generator of the waste. Under the protocol, the liability of the generator ends once the waste is transferred to the notifier and/or disposer. This, they argue, does not provide sufficient incentive to reduce hazardous waste generation and only encourages its export. Moreover, it is contrary to many national laws, particularly among OECD countries, that assign liability to waste generators rather than to disposal contractors. Further, the protocol covers only damage that occurs during shipment or initial deposit or processing and not damage that occurs afterward. This is problematic, as most damage from hazardous waste disposal and recycling manifests itself after disposal or processing has occurred, when sufficient residues and toxins have accumulated to pose serious health and environmental risks. Finally, environmental NGOs also complained that the protocol did not

---

[22] See Daniel Pruzin and Cheryl Hogue, "Negotiations Stall on Basel Protocol on Liability, Compensation for Spills," *International Environment Reporter* 22, no. 19 (September 15, 1999): 747; Daniel Pruzin, "Basel Protocol on Liability, Compensation Formally Opens in Switzerland for Signature," *International Environment Reporter* 23, no. 6 (March 15, 2000): 255.

set adequate minimum financial limits for liability in case of damage, though this issue will be revisited at the next COP meeting.[23]

The liability protocol to the Basel Convention has some aspects that weaken its effectiveness. Environmental NGOs argued that it would have been better to delay the adoption of the protocol in order to secure a stronger agreement. Remedying these weaknesses will be more difficult, but not impossible, now that the protocol has been adopted. It is important that these improvements to the protocol be considered. But this agreement covered only transboundary shipments of hazardous waste. There is also still scope for designing a global liability regime that applies to waste management generally, regardless of whether it is shipped from one country to another. A key provision would need to be the assignment of liability squarely on the generator. Without this, there is little incentive to reduce hazardous waste generation through the adoption of clean production technologies. This would avoid many of the problems with the Basel liability protocol, especially if it covered all disposal and treatment operations, regardless of where they occur and regardless of the origin of the firm responsible. This of course would require some degree of harmonization of national liability rules, or at least the adoption of some sort of "template" for such laws, especially for countries that do not currently have environmental liability laws. A global agreement on this matter would be extremely complicated to negotiate and implement. While it may be politically difficult if not impossible to achieve, a harmonized system of liability for hazardous waste would go a long way toward discouraging its generation.

Active Promotion of Clean Production

It is imperative that governments strengthen their commitment to the development and adoption of clean production methods both within and outside of their borders. To do this, there will need to be a clear definition of clean production and its relationship to environmentally sound management of wastes. There will also need to be support for the development and promotion of clean technologies, as well as stronger national

---

[23] For detailed NGO analyses of the protocol, see *BAN Report and Analysis of the Fifth Conference of Parties to the Basel Convention*, December 6–10, 1999; BAN, *Report of the 10th Meeting of the Ad Hoc Working Group of Legal and Technical Experts to Consider and Develop a Draft Protocol on Liability and Compensation for Damage Resulting from Transboundary Movements of Hazardous Wastes and Their Disposal*, August 30–September 3, Geneva, 1999; BAN, *Saving the Basel Liability Protocol*, August 30–September 3, 1999.

laws regarding the management of hazardous wastes and the promotion of clean production methods.

There are currently differences in the way various actors use and understand terms such as "clean production" and "environmentally sound management." Universal definitions must be adopted. In promoting a reduction in the generation of hazardous wastes, the Basel Convention has focused more on environmentally sound management than on clean production per se, though it does not see these goals as mutually exclusive. The Basel Convention initially defined ESM very broadly as management of wastes in a way that protects human health and the environment. The recently adopted Basel Declaration on Environmentally Sound Management takes a step closer to defining its relationship to clean production. It states that the ministers must "assert a vision that the environmentally sound management of hazardous and other wastes is accessible to all Parties, emphasizing the minimization of such wastes and the strengthening of capacity-building."[24] It also states that environmentally sound management efforts should be made with respect to prevention, minimization, recycling, recovery and disposal of hazardous wastes, as well as the promotion and use of cleaner production technologies.[25]

While this move is welcomed by many, some are concerned that the process of bringing clean production more fully into the scope of the convention's activities through the declaration on ESM may in fact weaken the concept. The Basel Action Network, for example, is worried that the mention of recycling and disposal alongside minimization as components of ESM appears to give these methods of addressing the problem equal weight.[26] According to BAN, if the Basel definition of ESM gives equal priority to these methods, then waste minimization via clean production loses the special emphasis that it should have. BAN's concern about this issue initially arose because the agenda of an OECD workshop held in October 1999 on the theme of ESM for wastes destined for recycling omitted any reference to waste minimization and prevention or the Basel Ban Amendment. When queried about this omission, the OECD responded that the workshop was on ESM, not on the Basel ban, waste minimization, or clean production. The OECD viewed the workshop as solely working

---

[24] Secretariat of the Basel Convention, "Basel Declaration on Environmentally Sound Management," *Report of the Fifth Meeting of the Conference of the Parties to the Basel Convention*, UNEP/CHW.5/29, December 10, 1999, 85–86.
[25] Ibid.
[26] BAN, *BAN Report and Analysis of the Fifth Conference of Parties to the Basel Convention*, December 6–10, 1999.

toward the development of criteria for the environmentally sound management of hazardous waste destined for recycling. A top OECD official defended this position, saying of the environmental NGOs, "They're just lumping everything together there.... Clean production is an issue in itself, as is waste minimization. But this meeting was planned to talk about ESM, and we think it was most effective to focus on the specifics of that issue."[27] This response prompted a coalition of environmental NGOs to boycott the OECD workshop.[28] Environmental NGOs thus expressed concern over the actual implementation of the Basel declaration, and they are especially concerned about which activities listed under the heading of ESM are allocated funds.

This debate over ESM and its relationship to clean production will have to be resolved at the international and national levels. If it is not, there is the risk that the focus on ESM will overshadow and exclude the goal of clean production. The conference of parties adopted a provisional work schedule for the years 2000–2002 for implementing the Declaration on Environmentally Sound Management. This includes a proposal for a second conference in Dakar, to further define the concept of ESM and to enhance partnership with all stakeholders. Environmental NGOs are concerned about this process because this conference is going to be funded by countries that oppose the Basel ban, namely the United States and Australia, and will likely also be subject to strong influence from industry groups.[29] They are also concerned because the secretariat of the Basel Convention has planned a "confidence building" meeting with industry groups, presumably in an attempt to attract funding from industry for the upcoming Dakar Conference.[30] This may lead to a strong industry influence over the agenda of that conference, while environmental NGOs have not been given a similar standing in the process. It is important that all stakeholders participate in this conference to define ESM and its relationship to clean production.

---

[27] Lawrence Speer, "NGOs Refuse to Participate in OECD Talks on Sound Strategies for Managing Waste," *International Environment Reporter* 22, no. 23 (November 10, 1999): 919.
[28] Jim Puckett, Open Letter Announcing NGO Boycott of OECD Workshop on Environmentally Sound Management of Waste Recycling, October 25, 1999, on behalf of BAN, Greenpeace International, Clean Production Action, and the Northern Alliance for Sustainability (ANPED), www.ban.org/Library/oecd_let.html; see also Speer 1999, "NGOs Refuse to Participate in OECD Talks," 919.
[29] BAN, *BAN Report and Analysis of the Fifth Conference of the Parties to the Basel Convention*, December 6–10, 1999.
[30] BAN, *Report on the Basel Convention's 17th Technical Working Group (TWG) and 2nd Legal Working Group (LWG)*, Geneva, October 9–13, 2000.

Public research and support for technology development is another activity that will be important in the promotion of clean production. This type of public research support for clean production technologies will need to pay particular attention to its applicability for and transfer to developing countries. The Basel Convention, through its establishment of regional centers for training and technology transfer, has promoted this sort of activity. As of late 1999 some twelve centers were in the process of being established. These centers are to focus on capacity building by providing technical guidance and advice on enforcement of the convention. The centers were also given the mandate to promote adoption of cleaner production technology adoption and environmentally sound hazardous waste management. The main problem plaguing these centers, however, is lack of adequate funding. The parties to the convention have provided seed funding to get them off the ground, but a long-term financial mechanism is yet to be worked out.[31]

Other efforts along these lines are also under way. Over the past few years a growing number of organizations have become devoted to promoting cleaner production, including industry organizations, intergovernmental bodies, and environmental NGOs. The UNEP Division of Technology, Industry, and Economics (DTIE) runs a Cleaner Production Program that promotes the adoption of cleaner technology and aims to reach a global consensus vision on cleaner production. It acts as a liaison between the various groups working to promote cleaner production. Its links, however, appear to be closest with industry groups. The DTIE's Cleaner Production Program also undertakes its own projects for development and dissemination of cleaner technology. To reinforce this work it has set up national centers on cleaner production in several developing countries to demonstrate these technologies. In 1998 it launched an international declaration on cleaner production, a voluntary declaration with an aim to bring together business, industry, NGOs, governments, and communities to commit to cleaner production.[32] As of November 2000, some forty-two governments, sixty-one businesses, fifty industry associations, eight NGOs, and various others had signed on. The declaration sets out important goals for industry, but to date it has not received the high profile or promotion that it deserves. There also appears to be little coordination between the UNEP Cleaner Production Program and the Basel Convention.

---

[31] Secretariat of the Basel Convention, "Regional Centres," http:www.basel.int/centers.
[32] UNEP DTIE, "Cleaner Production," http://www.uneptie.org/Cp2.

These efforts are commendable but are unfortunately lacking in sufficient funds and profile to make a strong impact on industry. Additional financial assistance to promote clean technology development and dissemination will be necessary and will most likely have to come from national or global bodies such as the World Bank and the Global Environment Facility. But it would be necessary to ensure that such assistance is free from the current ties to and bias toward cleanup as opposed to clean production technologies. It will also be important to ensure that publicly funded export credit agencies also follow clean production goals.

In addition to these international efforts, it is important that governments, rich and poor alike, enact stricter regulations on firms regarding emissions controls and waste management and that they develop policies to encourage adoption of clean production technology. Though this may be unpopular among governments and firms, a major survey of TNCs has shown that the primary motivator for firms to improve environmental practice has been government regulations.[33] Command and control (CAC) measures have been discredited in this era of liberalization and rolling back of the state as overly bureaucratic and inefficient. But some are arguing that CAC has been cast aside too soon and should be given a second look because in some cases it is as efficient if not more so than voluntary measures in achieving desired environmental outcomes.[34]

It is possible to develop strong government regulatory programs for hazardous waste management in both rich and poor countries. Such a measure would likely raise environmental costs in the developing countries to levels that approach those in the industrialized world, thus closing the gap in cost differentials. But as emphasized in a recent report on comparing the development of hazardous waste programs in eight countries, it takes at least ten to fifteen years to get a proper hazardous waste management system up and running to the point that firms adopt a culture of compliance.[35] It may take longer than this to move to a culture of clean production. Some rich industrialized countries have successfully encouraged waste minimization and recycling through nonmarket measures such

---

[33] UNCTAD, Programme on TNCs, *Environmental Management in Transnational Corporations: Report on the Benchmark Corporate Environmental Survey* (New York: United Nations, 1993), 38.

[34] Daniel Cole and Peter Grossman, "When Is Command-and-Control Efficient? Institutions, Technology, and the Comparative Efficiency of Alternative Regulatory Regimes for Environmental Protection," *Wisconsin Law Review* 1999, no. 5 (1999): 887–938.

[35] Katherine Probst and Thomas Beierle, *The Evolution of Hazardous Waste Programs: Lessons from Eight Countries* (Washington, D.C.: Resources for the Future, 1999).

as pollution prevention laws, public information available through pollutant release and transfer registers, strong liability legislation that places responsibility on the generator, and subsidies for acquisition of cleaner production technology.[36] These measures can be strengthened and incorporated into waste management programs in developing countries. The Basel Technical Working Group has already identified these measures for further study in its work plan for the Declaration on Environmentally Sound Management.

Pollutant release and transfer registers (PRTRs) in particular have great potential to encourage firms to adopt cleaner production methods. These registers require that firms make information available to the public regarding pollution releases. This type of register was recommended in Chapter 19 of Agenda 21. PRTRs are already in place in several countries, including the United States and Canada. The OECD sees PRTRs as a positive force for promoting toxic emissions reductions and has encouraged its member countries to adopt such programs, providing guidance on how to establish them.[37] With the assistance of international lending agencies, PRTRs are also being experimented with in some developing countries, including the Philippines and Indonesia.[38]

Stronger government regulations to promote clean production, particularly in developing countries, will likely be more effective if they are somewhat harmonized across countries. Though it is often argued that all countries have their own priorities and political systems and should thus develop their own particular environmental regulations, there may be strong benefits to some degree of harmonization. If the goal is universal adoption of clean production practices, it may be that some sort of model legislation, or template for regulatory structure can be developed based on successful cases of clean production promotion. This could then be used as a starting point for the adoption of similar rules, particularly in developing countries. Though the idea of harmonizing environmental regulations at the global level, at least with respect to hazardous waste management and clean production promotion may seem radical to some, it is not without

---

[36] Ibid., 49–50.
[37] OECD, *PRTR Implementation: Member Country Progress* (Paris: OECD, 2000), ENV/EPOC(2000)8/FINAL. See also "OECD's Work on Pollutant Release and Transfer Registers (PRTRs)," http://www.oecd.org/ehs/prtr/index.htm.
[38] "World Bank Endorses Disclosure of Emissions Data as Enforcement Technique," *International Environment Reporter* 19, no. 18 (September 4, 1996): 774–75; Rhea Sandique, "Rating System for 2000 Industries in Manila Set," *Manila Standard*, December 9, 1996; "Computer to List Firms Polluting Environment," *Philippine Daily Inquirer*, April 29, 1997.

precedent, particularly at the regional level. For example, this type of harmonization is a goal of the member states of the European Union.[39] And efforts along these lines are already under way in some of the rapidly industrializing countries in Southeast Asia.[40] These countries are also working at the regional level, through ASEAN, to strengthen national level waste disposal and waste trade laws and to set up an information exchange on the Internet for member countries in an attempt to keep out undesirable industries and wastes.[41] Moreover, this sort of harmonization is seen to be important among different jurisdictions within countries.[42] Its application at the global level with respect to hazardous waste management and clean production promotion is thus entirely possible and will likely be more effective than a myriad of different regulations that create incentives for hazards to relocate.

GLOBAL REGULATORY FRAMEWORK FOR TNCS

Corporate willingness to adopt clean production as the norm for industrial production will be an essential component of the effort to halt hazard transfer. Though there is a great deal of talk about "corporate greening," in practice, only a few leading firms have actually made important strides toward cleaner production. Some in the business and the environment literature have been pushing for a change in corporate culture toward cleaner production.[43] They argue that it is in industry's best economic interest to pursue such a strategy, mainly because of economic efficiency gains over the long run. The measures that I have outlined above involve the application of outside pressure and incentives for firms to adopt cleaner production. Some firms have responded to this pressure with change from within that meets this goal.[44] Some key firms in the chemical industry have even

---

[39] Kate O'Neill, *Waste Trading among Rich Nations: Building a New Theory of Environmental Regulation* (Cambridge, Mass.: MIT Press, 2000); Kate O'Neill, "The Changing Nature of Global Waste Management for the 21st C.: A Mixed Blessing?" *Global Environmental Politics* 1, no. 1 (2001).

[40] Probst and Beierle 1999.

[41] "ASEAN Calls for Regional Clampdown on Illegal Hazardous Waste Movements," *International Environment Reporter* 23, no. 16 (August 2, 2000): 600.

[42] Probst and Beierle 1999, 23–24.

[43] For example, Stephan Schmidheiny with WBCSD, *Changing Course: A Global Business Perspective on Development and the Environment* (Cambridge, Mass.: MIT Press, 1992); Livio DeSimone and Frank Popoff with WBCSD, *Eco-Efficiency: The Business Link to Sustainable Development* (Cambridge, Mass.: MIT Press, 1997).

[44] On the question of corporate decision making for environmental improvement, see Aseem Prakash, *Greening the Firm: The Politics of Corporate Environmentalism* (Cambridge: Cambridge University Press, 2000).

taken this new green outlook to their branches and affiliates in developing countries via national level voluntary standards such as the American Chemistry Council's Responsible Care program.[45] But the culture of clean production needs to be widespread, rather than being adopted by only a few leading firms. If the recent past is any guide, it is not likely that firms will adopt cleaner production technologies on their own in the short run. Until they do, other shorter run measures will also be necessary.

A variety of measures could be taken at the global level to encourage TNCs to adopt cleaner production methods. In particular, a binding global agreement could stipulate environmental criteria to be followed by TNCs. In its strongest form, this could include performance-based criteria with respect to hazardous waste management and clean production, or at the very least it could require firms to abide by the environmental regulations in their home countries, as recommended in Agenda 21. Such an agreement could also require TNCs to disclose publicly information about hazardous waste generation similar to the national PRTRs and transfer registers. This informationwould allow developing country governments and local communities to screen foreign investments before deciding whether to accept them.[46] Thus far few transnational firms have been willing to provide this information voluntarily. The availability of such information on environmental performance of TNCs could be very effective, especially if it was combined with a global liability agreement that places the responsibility of hazardous waste on the generator. These measures could help to prevent TNCs from taking advantage of regulatory differences between countries as they phase in stronger and more harmonized hazardous waste management and clean production regulations. Advocating this type of global regulatory framework does not preclude voluntary and market-based initiatives already endorsed by firms that ostensibly have similar goals. It simply recognizes that regulations imposed from outside can and do matter to the inside workings of firms.

Though the recommendation of a global set of environmental requirements for TNCs may seem somewhat extreme to some, the beginning of such an agreement is already well under way within the OECD. In July 2000, the OECD substantially revised its Guidelines for Multinational Enterprises to include recommendations for environmental practices for

---

[45] Ronie Garcia-Johnson, *Exporting Environmentalism: U.S. Multinational Chemical Corporations in Brazil and Mexico* (Cambridge, Mass.: MIT Press, 2000).
[46] See, for example, United Nations Centre on Transnational Corporations, *Transnational Corporations and Industrial Hazards Disclosure* (New York: United Nations, 1991).

firms.[47] The new guidelines promote already existing environmental management standards such as the ISO 14000. But they also call for TNCs to adopt "measurable objectives, and where appropriate, targets for improved environmental performance."[48] By stressing environmental performance, the guidelines are stronger than ISO 14000 requirements. The guidelines also call for improved availability of information on the environmental activities of firms, as well as consultation with affected communities. Though these guidelines are not binding and apply at the present time only to OECD countries, they have the potential to form the basis for a global agreement, as has happened with OECD guidelines on other issues, including the trade in hazardous wastes.

Developing country governments, most of whom are desperate for foreign investment, might resist the notion of making these guidelines binding and global. The main argument that has been aired against global rules for TNCs is that differentials in environmental regulations and associated costs are a "natural" part of developing countries' comparative advantage and would discourage foreign investment, especially if TNCs are targeted for tighter regulations at the global level. Mexico, for example, one of the poorer members of the OECD, strongly resisted the inclusion of a strong compliance mechanism in the OECD Guidelines for Multinational Enterprises. It stressed that it did not want environmental NGOs using the guidelines to "harass" firms investing in Mexico.[49] But if the goal is the global application of clean production technologies, such requirements would work toward improving environmental performance, at least among TNCs, operating in developing countries.

NGO Pressure

Continued NGO pressure with respect to all of the above measures will be vital in the effort to stop hazard transfer, as it has been in the past. The work of Greenpeace was instrumental in the adoption of the Basel Ban Decision and Amendment. Similarly, the work of BAN has been extremely important in exposing cases of hazard transfer that breach the ban, and in promoting government ratification of the ban amendment. BAN also pushed for the emphasis on waste minimization in the Basel Declaration.

---

[47] These guidelines were first adopted in 1976 and were revised in 1991 and in 2000.
[48] OECD, *The OECD Guidelines for Multinational Enterprises*, http:/www.oecd.org//daf/investment/guidelines/mnetest.htm.
[49] Lawrence Speer, "New Guidelines on Business Behavior, Environment Approved by OECD Nations," *International Environment Reporter* 23, no. 14 (July 5, 2000): 519.

Without these efforts, the global framework for regulating hazard transfer would likely be much weaker than it currently is. But the forces that are attempting to weaken that framework, primarily the efforts on the part of certain industry groups and key industrialized country governments to reverse the Basel ban, are still strong. In this context, it is important that NGOs keep up their pressure.

The work ahead for environmental NGOs will be extremely challenging. In some ways these groups may be victims of their own success thus far. With the de facto ban in place and the reduction in the number of scandalous cases of waste trade between rich and poor countries, the sense of urgency on the part of governments to ratify the ban amendment has diminished as well. The environmental NGOs fear that if the ban were to be reversed or severely weakened, hazard transfer from rich to poor countries will not only resurface, but also become the norm. The factors that encourage hazard transfer in the first place—the growth in hazardous waste generation and rising costs of disposing it, economic globalization, and the rich-poor divide—are today stronger than ever. Environmental groups feel that the rules that have been put in place have held back much of the worst of hazard transfer. If they are removed, the situation could rapidly deteriorate.[50] But convincing governments and the public that this is the case is extremely difficult when there are fewer cases of waste trade to point to. It is also difficult to document the reasons behind the growth in hazardous investment in developing countries with the same degree of clarity as was possible with the early waste trade cases. And it is difficult to demonstrate the potentially large benefits of cleaner production when at present it is so little practiced. Garnering support for their case to halt the waste trade in the late 1980s and early 1990s may have been much easier because it was a blatant and growing problem, and because it clearly contravened people's sense of justice. Convincing people of a potential negative scenario that needs to be avoided, and of the as yet to be realized benefits of wide scale adoption of cleaner production, is likely to be much harder. But it is nevertheless a vital task.

Environmental NGOs know they will need to continue efforts to maintain the de facto ban that is currently in place, and to solidify its legal status through the promotion of ratification of the ban amendment. BAN's campaign on this front attempts to show that the global community has avoided a major crisis by adopting the ban before the problem became

---

[50] Jim Puckett, "The Basel Convention's Ban on Hazardous Waste Exports: An Unfinished Success Story," *International Environment Reporter* 23, no. 25 (December 6, 2000): 984.

completely out of control. They are using the waste trade cases that breach the ban to illustrate what the global situation could look like in the absence of these rules. It is their hope that this argument will convince governments to ratify the ban amendment. At the same time, environmental NGOs are continuing to directly fight efforts to weaken the ban amendment. They have continued to participate in the Basel Convention's Technical Working Group and Legal Working Group meetings, where industry groups and key industrialized country governments have been most active in their attempts to undermine the ban.

Environmental NGOs are also increasingly paying greater attention to the need to promote cleaner production in their work. Greenpeace, for example, has a clean production campaign that highlights the existence of clean production technologies and their benefits. It has funded projects for the development and dissemination of clean production technologies. BAN has also been working in this area. It continues to demand that ESM, as defined in the Basel Convention and elsewhere, emphasize waste minimization and avoidance, and not recycling and disposal. Their strategy has been to argue that so long as recycling and disposal are seen as legitimate activities with respect to hazardous waste, there will be little incentive to minimize waste generation with cleaner production technologies. There are new groups emerging, such as Clean Production Action, which make clean production a central focus of their work. It will be important for these groups to continue to expand their work, as well as to find ways to collaborate with the work of the UNEP Cleaner Production Program and the more business-oriented groups pushing for cleaner production.

Conclusion

The dynamic nature of the hazard transfer problem has made tackling it at the global level particularly difficult. As the above recommendations indicate, addressing the problem requires measures that cover a range of issues, from the trade in wastes to the promotion of clean production. These measures must also be flexible to adapt to the changing nature of the problem. They will also work effectively only if they are pursued together and with vigor. The political commitment to achieve the required measures for stopping hazard transfer will come only once the project of clean production is fully embraced by firms, governments, environmental NGOs, and consumers in both rich and poor countries. Without this commitment, the problem of hazard transfer will likely find new outlets in response to partial and weak regulations, much as we have already seen. The recommendations I have laid out in this chapter have rested on the assumption that global production will need to continue to grow in the future. An embrace of cleaner production technologies can help to reduce the rate of growth of the mountains of hazardous wastes that are generated each year. Another effective way to reduce that growth, though much more controversial, is to slow the pace of industrial production itself, and in turn, levels of consumption. This, in conjunction with the adoption of cleaner production technologies, could go a long way toward solving the problems of hazardous waste generation more broadly and hazard transfer in particular.

Index

Addis Ababa, 41
Africa, 32, 34, 62
   role in negotiation of Basel Convention, 41–43, 46
African, Caribbean, and Pacific (ACP) states, 48, 60
African Ministerial Conference (Dakar), 42
Agenda 21, 128, 131, 137, 138, 142, 144, 168
Alang Shipyard, India, 101
Albania, 66
Alter, Harvey, 75–76, 85, 89, 94
American Chemistry Council, 168
Angola, 62, 121
Annex I of the Basel Convention, 44, 83
Annex III of the Basel Convention, 44, 83
Annex VII of the Basel Convention, 77, 82, 87, 93–94, 98–100, 101–103, 116
Annex VIII of the Basel Convention, 96–97, 99–100, 157
Annex IX of the Basel Convention, 96, 99–100, 157
Antarctica, 45
Argentina, 97, 144
arsenic, 153
Article 6 of the Basel Convention, 45
Article 11 of the Basel Convention, 45, 57, 60, 77, 82, 87, 94, 98–100, 103, 160
Article 12 of the Basel Convention, 100
Article 13 of the Basel Convention, 158
Article 16 of the Basel Convention, 158
asbestos, 24, 35, 101
Asia, 32, 116–117
Asian Development Bank, 67
Asian financial crisis, 134
Asia-Pacific, 132, 135

Association of Southeast Asian Nations (ASEAN), 51, 167
Australia, 50, 55, 65, 68, 71–73, 76, 153
Awori, Achoka, 49

Bahamas, 33
Bamako Convention, 48–49, 55, 58, 63
Bangladesh, 67
Barcelona Convention, 50
Basel Action Network (BAN), 82, 92–93, 96–101, 152–153, 162–163, 169–171
Basel Ban Amendment, 116, 151
   adoption of, 76–80
   industry's arguments against, 86–90
   ratification of, 4, 90–92, 99, 152, 157–159, 170–171
   See also Decision III/1 of the Basel Convention
Basel Convention, 2, 3
   legal weaknesses of, 53–58
   negotiation of, 38–47
   negotiation of waste trade ban, 67–80
   parties to, 156–157
   provisions of, 44–47
   trade provisions, 45–46
   See also specific articles, decisions, COP meetings and subgroups
Basel Declaration on Environmentally Sound Management, 102, 156, 162–163, 166, 169
Bayer, 117, 119
benzedine dyes, 107
Bhopal, India, 119
bilateral waste trade agreements. *See* Article 11 of the Basel Convention
Brazil, 65, 76, 97, 101, 118, 158

British Environmental Management Standard BS 7750, 138
Bullock, John, 85, 88
Bureau of International Recycling (BIR), 84, 86–88, 90, 93–96
Bush, George Sr., 55
Business Charter for Sustainable Development (ICC), 136
business/industry groups, 2–4, 15–19, 151, 156, 163
   in ISO 14000 process, 147
   recycling industry, 81–90, 93–102
   role in negotiation of Basel Convention, 40, 43
   role in politics of the Basel ban, 81–103
Business Recycling Council, 85, 99

cadmium, 61, 67
California, 115
Cambodia, 154–155
Canada, 55, 65, 68–69, 71–73, 76, 97, 132, 139, 153, 166
Caribbean, 32–33
Caribbean Community (CARICOM), 42
Cascio, Joseph, 142–143
Cato Ridge, South Africa, 62–64
Central America, regional waste trade agreement, 49–50
Central Europe, 32–33, 68–69
CERES Principles (Coalition for Environmentally Responsible Economies), 136
chemical industry, 16, 107, 114, 116–118, 120–122, 167
chemicals, 37
Chile, 97
China, 64, 71, 120, 124
chloralkalai, 117–118, 120
chlorine industry, 117–118, 120
clean/cleaner production, 2, 4–5, 20, 127
   promotion of, 155, 161–167
Cleaner Production Program (UNEP), 128, 164, 171
Clean Production Action, 171
clean production technologies, 121–124, 126–130, 150, 161
   costs of, 129–130, 132
   transfer of, 19, 127–130, 141–143, 145, 147–148, 164
cleanup technologies, 5, 130–136, 147
Clinton, Bill, 55
command and control measures, 165
comparative advantage, 6

compliance with Basel Convention, 57, 159
Conakry, Guinea, 36
Conference of the parties (COP) meetings of Basel Convention, 46
   COP-1, 68–70
   COP-2, 70–76, 85–86
   COP-3, 76–79, 84, 86, 90, 93
   COP-4, 84, 94, 96–99, 157
   COP-5, 99–102, 157–158, 160
Consultative Sub-Group of Legal and Technical Experts of the Basel Convention, 98, 100
consumption, 171
copper
   cables, 153
   smelter furnace dust, 67
   smelting, 107
Cuba, 145

Dakar, Senegal, 42, 75–76, 163
Dansk Sojakagen Industries, 120
data on hazardous waste trade, 24–31, 34, 58
Dauvergne, Peter, 16
DDT, 37, 66
debt, international, 11, 23
Decision II/12 of the Basel Convention, 73–77, 91, 101, 151
Decision III/1 of the Basel Convention, 77–78, 91, 101, 151
definition of hazardous waste, 4, 19, 24, 44, 56, 83, 88, 93–98
Denmark, 66, 68, 120
Developing Countries (DEVCO) assistance program of ISO, 146
dioxin, 24, 117, 152
Division of Technology Industry and Economics (DTIE), 164
double standards, 112–113, 118–119
Dow, 131
Dowdswell, Elizabeth, 73
DuPont, 110, 119, 131
Durban, South Africa, 119

Earthlife Africa, 63
East Asia, 34
Eastern Europe, 27, 32–33, 61–62, 68–69, 76
ecoefficiency, 127
Economic Community of Latin American Countries (ECLAC), 51

Economic Community of West African States (ECOWAS), 42
economic globalization, 2, 6–7, 10–14, 21–24, 51, 79, 111–112, 122, 125, 127, 148, 151, 170
economic growth, 6–7, 108
Egypt, 62
electronics industry, 16, 105, 114, 122
El Tawil, Anwar, 146n
end of the pipe technologies, 128, 130
environmental cost of firms, 110–112, 121, 125
environmental health and safety policies, 113
Environmental Justice and Networking Forum, 63
Environmental Management and Audit Scheme (EMAS), 138
environmental management systems (EMS) standards, 19, 138. *See also* ISO 14000
environmental NGOs, 2, 3, 5, 14–18, 20, 151
  alliance with Third World Countries, 43, 52, 70
  clean production strategy, 162, 169–171
  role in evolution of Basel Ban Amendment, 67–80
  role in negotiation of Basel Convention, 38–51
  role in post-Basel Ban Amendment politics, 91–93, 95–97
  role in setting of ISO 14000 standards, 139
environmental performance standards, 137, 139, 141, 147
environmental regulations
  enforcement of, 110–111, 114, 118, 123, 134–135, 137, 140, 144
  stringency of, 105, 108, 111, 114–116, 133, 143–144, 165, 168
environmental services market, 130–135
environmentally sound management (ESM), 56, 127, 156
  definition of, 162–163, 171
environmentally sound technologies, 127. *See also* clean production technologies
Estonia, 69
Europe, 24, 34, 116, 134
European Commission, 86
European Community (EC), 47, 48
  waste trade regulations, 39
  See also European Union (EU)
European Union (EU), 48, 71–73, 75–76, 84, 96–98, 157, 167
  waste trade regulations, 60–61
export credit agencies, 121, 165
Eyadema, Gnassingbe, 41

fertilizer, 64–67
Finland, 68, 146
Florida, 34
Formosa Plastics Group (FPG), 154
France, 47, 154
Frey, Scott, 115
Friends of the Earth International, 88
furans, 117
furniture industry, 114–115

GATT, 58, 143. *See also* World Trade Organization (WTO)
Germany, 55, 64–66, 68–69, 71–73, 153–154
global civil society, 13, 14. *See also* nonstate actors
Global Environment Facility, 165
global warming, 36
Goa, India, 119
Gore, Al, 55
Great Britain, 101. *See also* United Kingdom
Greenpeace, 32, 50, 55, 61, 97, 120
  clean production campaign, 171
  negotiation of Basel Ban Amendment, 69–170
  relationship with industry groups, 85
  role in negotiation of Bamako Convention, 48–49
  role in negotiation of Basel Convention, 43, 45, 47
  waste trade campaign, 39, 63, 82, 91–93, 151, 153, 169
Group of 77 (G-77), 43, 46, 68–69, 71–72, 80
Guinea, 36
Guinea-Bissau, 35

Hadlock, Charles, 110
Haiti, 33–34
Hall, Derek, 107n
hazardous industry, 2, 9–10, 12, 104–125, 150
  role of TNCs, 112–121
hazardous plant equipment, 112, 119–120
hazardous waste
  disposal costs, 23, 108, 170
  disposal facilities, 133–135

Index   175

hazardous waste *(continued)*
  generation of, 24–31, 45
  See also specific aspects of hazardous waste and hazardous waste trade
HAZTRAKS, 115
health implications of hazardous waste, 24, 26, 64, 123
heavy metals, 24, 107
hexachlorobenzine (HBC), 118
Hirschman, Albert O., 87n
Hong Kong, 133–134, 153
Horne, Scott, 87
Hungary, 39, 68

illegal waste transfers, 31, 46, 56, 67, 89, 152, 158–159
incineration, 134, 152
  of hazardous wastes, 61–62, 66
India, 64, 76, 101, 119, 120, 153–154, 158
Indo Era Multa Logam (IMLI) battery recycling plant, 65
Indonesia, 64–66, 120, 124, 134, 166
industrial flight hypothesis, 106
industrial location, 8–10, 19, 104–112, 115–116
industrial zoning policies, 123
Institute of Scrap Recycling Industries (ISRI), 85, 87–88
International Chamber of Commerce (ICC), 40, 75–76, 85, 90, 95
  Business Charter for Sustainable Development, 136
International Council on Metals and the Environment (ICME), 85, 87, 89–90, 99
International Finance Corporation (World Bank), 135
International Maritime Organization, 102
International Monetary Fund, 11, 111
International Organization for Standardization (ISO), 136–140, 144–147
  standard setting procedures, 138, 144–147
International Precious Metals Institute (IPMI), 40, 85
International Toxics Investigator, 43n, 91
INTERPOL, 159
ISO 14000 environmental management standards, 5, 18–19, 136–149, 169
  certification procedures, 139–142
  role in developing countries in, 145–148
  role of NGOs in, 146–148
  role of TNCs in, 145–148
  standard-setting process, 144–147
Israel, 97, 99
Italy, 23, 35, 47, 68
Izmir Protocol, 50

Japan, 47, 68–69, 71, 73, 76, 107, 135, 152

Kante, Bakary, 93
Karin B, 35
Karliner, Joshua, 132
Kassa Island, Guinea, 36
Keck, Margaret, 13
Khian Sea, 2, 33
Koko, Nigeria, 35
Krueger, Jonathan, 49n, 159n
Kummer, Katharina, 39n, 50, 157

landfills, 23, 32, 35–36, 117, 122
La Paz Agreement, 114
Latin America, 32, 62, 116
Latvia, 64
lead, 61, 67, 101, 153, 158
lead-acid batteries (recycling of), 64–66, 153
leather tanning, 107, 122, 124
Legal Working Group of the Basel Convention, 102, 159, 171
Leonard, H. Jeffrey, 106–107
liability, 57
  assignment of, 160–161
  costs of, 109–110, 130, 131–132
  See also Protocol on Liability and Compensation (Basel Convention)
Lipschutz, Ronnie, 13
List A wastes in the Basel Convention, 94–97
List B wastes in the Basel Convention, 94–97
Lomé Convention (Lomé IV), 48, 58, 61, 63
Lomé, Togo, 42
Love Canal, 23
Lutzenberger, Jose, 2

Malaysia, 118, 133–134
Managua, Nicaragua, 117
Mani, Muthukumara, 107–108
maquiladora firms, 114–116
market-based initiatives, 19
Marshall Islands, 36
Mediterranean, regional waste trade agreement, 50
mercury, 62–64, 66, 101, 117–118, 120, 154

metal plating, 122
Mexico, 35, 65, 101, 114–116, 121, 134, 169
Middle East, 34
mislabeling of hazardous waste, 54
Mitsubishi, 118
Monaco, 97
Monsanto, 131
Montreal Protocol, 17, 117
Mozambique, 66
multilateral environmental agreements (MEAs), 7–8, 10, 144–145, 148

Namibia, 62
national waste trade legislation, 47–48
Netherlands, 146
newly industrializing countries, 110, 116, 132
New York City, 62
New Zealand, 50, 65
Niagara Falls, New York, 118
Nicaragua, 117
Nigeria, 2, 35, 64
NIMBY (not in my backyard) syndrome, 23
Non-Aligned Movement, 42
nonstate actors, 2, 12–18, 22, 38, 51–52, 94, 103, 151. *See also* business/industry groups; environmental NGOs
Norske Skog, 120
North American Free Trade Agreement (NAFTA), 115–116
Norway, 36, 47, 68, 120

OECD, 3, 24–25, 27, 70, 116
    clean production efforts, 166
    data on OECD to non-OECD waste trade, 27–30, 58–59
    data on waste generation, 25
    guidelines for multinational enterprises, 168–169
    role in negotiation of Basel Ban Decision and Amendment, 68–80
    role in negotiation of Basel Convention and COPs, 38–40
    waste trade regulations, 39–40, 60–61
    workshop on environmentally sound management, 162–163
Organization of African Unity (OAU), 41, 46, 48, 62
Oslo, Norway, 145

Pacific Rim, 116
Pakistan, 120

Papan, Malaysia, 118
Papua New Guinea, 50
PCBs, 24, 35, 101
Pennwalt Corporation, 117–118
pesticides, 37, 56, 64, 66, 107, 122
Philadelphia, Pennsylvania, 33
Philippine Recyclers Inc. (PRI), 65
Philippines, 64, 124, 152–153, 158, 166
plastic, 24
    recycling of, 64, 85
Poland, 68
pollutant release and transfer registers (PRTRs), 166, 168
pollution control costs, 106–107, 108–109
pollution haven hypothesis, 7–10, 106–108
pollution prevention, 127–128, 141, 166
Pomerance, Rafe, 88
Porter, Gareth, 10, 110–111
potassium cyanide, 133
prior notification, 45, 57–58, 60, 101
protests against hazardous industry, 107, 113
Protocol on Liability and Compensation (Basel Convention), 100, 159–161
Puckett, Jim, 92, 153, 157, 158n
PVC (polyvinyl chloride), 62, 117, 153

race to the bottom/top, 7–8, 10, 110
radioactive wastes, 44, 118
Ramcar Batteries, 65
Ravi Alkalis, 2–4
recycling of hazardous waste, 2–4, 18, 53–54, 58–67, 70, 76, 79–80, 150, 171
    See also specific types of hazardous waste
red, amber, green waste regulations, 60–61
regional waste trade agreements, 47–51
resource extraction, 105
responsible care, 168
Rhone-Poulenc, 118
Romania, 66
Russia, 69, 76

Saudi Arabia, 65, 153
Schneider, Manfred, 117
scrap metal, 83–84, 86–87, 89, 96, 153
Seattle, Washington, 14
Secretariat of the Basel Convention, 46, 57, 69, 72–73, 99, 158, 163
Senegal, 39, 75, 158
Seveso, Italy, 23
shipbreaking, 101–102
Sihanoukville, Cambodia, 154
Sikkink, Kathryn, 13

Singapore, 65, 153
Sklair, Leslie, 115
Slovenia, 97, 99
small and medium-scale enterprises (SMEs), 122–125
Somalia, 35
South Africa, 62–64, 97, 119, 145, 153
South America, 76
Southeast Asia, 76, 120, 122–124, 133, 167
South Korea, 76
South Pacific, 32
    regional waste trade agreement, 50
South Pacific Forum, 50
Spain, 101
Stairs, Kevin, 48n, 49n, 73n
Standards Council of Canada, 139
Strategic Advisory Group on the Environment (SAGE), 138
Strohm, Laura, 56n, 57n, 106n
structural adjustment programs, 11, 12
Sudan, 37
Summers, Lawrence, 1
Sweden, 68
Switzerland, 39, 68

Taiwan, 64, 101, 133, 154–155
Technical Barriers to Trade Agreement (of GATT), 143
Technical Committee 207 of ISO (TC 207), 139, 145–146
Technical Working Group of the Basel Convention, 77, 82, 84, 88, 93–103, 166, 171
textile industry, 84, 105, 122
Thailand, 65, 134
Thompson, Peter, 106n
Thor Chemicals, 62–64
Togo, 41, 42
Tolba, Mostafa, 36, 46–47, 55, 68–69, 157
Toxic Trade Update, 43n
trade and environment debates, 5–12
trade barriers, 143
trade competitiveness, 8–9
transnational corporations (TNCs), 4–5, 7–8, 11, 15–18, 61
    clean production initiatives, 126–130
    environmental standards of, 142
    investment in hazardous industry, 104–125
    regulatory framework for, 20, 167–169
    role in ISO 14000 process, 145–148
    See also specific corporations

Umgeni River, 63
UN Conference on Environment and Development (UNCED), 13, 15, 127–130, 138, 147
Union Carbide, 119
United Arab Emirates, 153
United Kingdom, 55, 62, 65, 68–69, 71–73, 153
United Nations Environment Program (UNEP), 24, 40, 42, 44, 55, 128, 164
United Phosphorus, 120
United States, 24, 34–35, 37, 47, 55, 62, 64–65, 67–69, 71–73, 76, 84, 96, 99, 109, 114–116, 131–132, 134–135, 142, 153–154, 156–157, 166
U.S. Agency for International Development (USAID), 36, 135
U.S.–Asia Environmental Partnership, 135
U.S. Chamber of Commerce, 75
U.S. Environmental Protection Agency (USEPA), 96
U.S.–Mexico Integrated Border Environmental Plan, 115

Veys, Francis, 86, 90
Voluntary environmental measures (industry), 17, 19, 128–129, 130, 136–137
    See also specific standards

Waigani Convention, 50
Wapner, Paul, 15
Waste Management International (WMX), 133–134
waste-to-energy schemes, 61–62
Waste Trade Update, 43n
Wheeler, David, 107–108
Wolfe, Thomas, 88
World Bank, 1, 11, 111, 124, 135, 147, 165
World Customs Organization, 159
World Health Organization (WHO), 63
World Trade Organization (WTO), 10, 83, 87, 89–90, 98–99, 143, 147
World Wide Fund for Nature, 146

Zimbabwe, 36
zinc, 61, 153, 158
Zone of Peace and Cooperation in the South Atlantic, 42